Theories of Inequality

Fifth Edition

Theories of Social Inequality

Fifth Edition

Edward G. Grabb
University of Western Ontario

NELSON / E D U C A T I O N

NELSON EDUCATION

Theories of Social Inequality
Fifth Edition
by Edward G. Grabb

Associate Vice President, Editorial Director:
Evelyn Veitch

Editor-in-Chief, Higher Education:
Anne Williams

Executive Editor:
Cara Yarzab

Executive Marketing Manager:
Kelly Smyth

Senior Developmental Editor:
Lesley Mann

Developmental Editor:
Sandra de Ruiter

Permissions Coordinator:
Kristiina Bowering

Content Production Manager:
Lara Caplan

Copy Editor/Proofreader:
Wendy Yano

Indexer:
Jin Tan

Production Coordinator:
Susan Ure

Design Director:
Ken Phipps

Interior Design:
Katherine Strain

Cover Design:
Johanna Liburd

Cover Image:
Sunday Afternoon on the Island of La Grande Jatte, 1884–86 (oil on canvas), Seurat, Georges Pierre (1859–91)/Art Institute of Chicago, IL, USA/Bridgeman Art Library

Compositor:
Interactive Composition Corporation

Library and Archives Canada Cataloguing in Publication Data

Grabb, Edward G.
Theories of social inequality/
Edward G. Grabb.—5th ed.

Includes bibliographical references and index.
ISBN 0-17-641666-8

1. Equality. 2. Power (Social sciences). 3. Social classes. I. Title.

HT609.G7 2006 305.5
C2006-902997-0

For Denise and Glenn

Contents

Chapter Three

Max Weber and the Multiple Bases for Inequality 34

Chapter Four
Durkheim, Social Solidarity, and Social Inequality 66

Chapter Five
Structural Functionalism and Social Inequality 88

Chapter Six
Recent Perspectives on Social Inequality 114

Chapter Seven
Theories of Social Inequality: An Overview and Evaluation 205

Preface

This is the fifth edition of *Theories of Social Inequality.* The principal goal of this volume is to review and assess the major conceptions of social inequality found in classical and contemporary sociological theory. The book will be of use to two somewhat different audiences: to students at various levels who can see the value in concise assessments of how leading theorists have dealt with the key conceptual problems in the field of inequality, and to colleagues who share my own interest in, and fascination with, the parallels and common ground to be found in a wide range of theoretical approaches, many of which have been viewed as essentially distinct or even contradictory.

The assistance and support of Thomson Nelson, along with the suggestions of many colleagues and students, have been very helpful once again. Based on their comments, some noteworthy changes have been made to the book. More examples and elaborations of key concepts have been included. In addition, because some of the contemporary theorists have continued to contribute significant works to the area since the last edition, I have enlarged the chapter on recent theorists to reflect these new developments. Generally, however, these changes and additions do not alter the fundamental goals of my analysis.

As ever, I have placed considerable emphasis on outlining and clarifying the ideas of Marx and Weber, whose works still provide the classical core for theoretical debates in the study of social inequality. Durkheim is included among the classical thinkers because of his unquestioned stature as a general theorist in sociology, his insightful and frequently overlooked observations on the particular issue of inequality, and his role as a bridge linking the early theorists to the more recent structural-functionalist approach. The structural-functionalist perspective, which falls in the middle of the chronology from early to recent theories, continues to be of importance in tracing the history of conceptual developments in the study of inequality. Although most observers no longer view this perspective as a serious alternative to the leading theoretical approaches, and most have been quite critical of its conservative implications and related difficulties, there has been renewed interest in structural functionalism in recent years. Moreover, several aspects of the structural-functionalist perspective are notable, both in their own right and because of the critical reactions they have stimulated among other theorists. This critical response marked the start of the subsequent movement back toward Marx and Weber and the renewed interest in debating and reworking the ideas of these two central figures.

The debate over the views of Marx and Weber still informs much of the current thinking in the study of social inequality. As a result, this debate and the key issues it has generated provide the principal focus for much of the material presented in the latter part of the book. For obvious reasons, it is not possible to conduct a thorough assessment of all the contemporary writers who have made a contribution to this literature. Following the earlier editions, I have concentrated on a detailed discussion of six recent theorists in the field: Dahrendorf, Lenski, Poulantzas, Wright, Parkin, and Giddens. The initial selection of these writers was guided by the belief, which I continue to hold, that their approaches represent a chronology of the key ideas that have predominated in most current views on how social inequality is to be conceived and understood. Therefore, although there are other writers who rival these six and could be examined in their place, I have retained the original set of recent theorists in this edition. Of course, other analysts are also considered at appropriate points in the presentation, even though their work is not assessed in as much detail.

It will be apparent in the book that my views on social inequality have a generally Weberian cast. Thus, although the concept of class, which is typically identified as a Marxian concern, is a central element in the analysis, it is the Weberian idea of power or domination that receives the greatest emphasis and that is used to tie the overall discussion together. This does not mean that I agree with those who now contend that class is a concept that has lost its significance in the modern era. My position is that class inequalities remain pivotal to any complete understanding of social inequality. In addition, however, I argue that class inequalities, like the other key forms of inequality that can be identified in social structures, should be conceived primarily as consequences of differential access to the major sources of power in society.

The recognition of these other forms of inequality is at the heart of a second point I wish to stress about my goals in this book. My choice of writers to review has been guided mainly by a search for general approaches to the study of inequality, as opposed to theories that focus exclusively on specific forms or bases of inequality, such as those involving race, gender, region, age, religion, and so on. I contend that the perspectives examined here are important precisely because they point us toward a more unified approach for analyzing all such forms of inequality. This approach centres, as I have already noted, on the problem of power or domination.

While I have not attempted a comprehensive exposition of perspectives dealing with specific types of inequality, theories that focus on problems of gender and race receive considerable attention. This reflects the trend among many recent theorists in the field of inequality to concentrate on these two important issues. I agree with those writers who argue that, in certain societies or social settings, inequalities based on gender or race are demonstrably more central than any other forms for understanding the overall pattern of inequality existing in those contexts. Again, however, I suggest that these specific types of inequality are like any other, in that they, too, are explicable essentially in terms of power differences among groups and the relations of domination that arise from such differences. I also make the argument, as outlined in Chapter 6, that there are some striking similarities and conceptual linkages between the more general classical and contemporary perspectives highlighted in the book and most leading theories of gender and race.

In this fifth edition of *Theories of Social Inequality,* I have also included a section in Chapter 6 that addresses the problem of social inequality at the transnational or global level. Interest in this problem is actually of long-standing, and can be found in the work of theorists that go back at least as far as Marx and Weber. Nevertheless, there is no doubt that the topic of globalization, including the crucial issue of explaining the often massive inequalities that currently exist among the world's nations, has received considerable attention in the social sciences over the past two decades. In keeping with the general theme of this book, here too I will attempt to show that global social inequality, like any other form, is best understood by examining the overarching influences of different forms of power both within and among nations.

I wish to thank a number of people for their help and suggestions during the preparation of the fifth edition. At Thomson Nelson, Sandra de Ruiter (developmental editor), Lesley Mann (senior developmental editor), and Cara Yarzab (executive editor) provided consistent and valuable editorial advice. Thanks too to Kristiina Bowering (photo and permissions research), Lara Caplan (content production manager), Kelly Smyth (executive marketing manager), and Wendy Yano (copy editor). I am also grateful for the facilities and other support made available to me by the Department of Sociology at the University of Western Ontario, the special and intellectually stimulating place where I have worked for some 30 years. There are more intellectual debts to acknowledge than I can list here. This edition's reviewers included Hans Bakker (University of Guelph), Jian Guan (University of Windsor), Richard Ogmundson (University of Victoria), and Lina Sunseri (University of Western Ontario). Their comments and those of the reviewers of earlier editions have been helpful in spurring me to rethink and rework the book in a number of ways. I have also received valuable suggestions and comments from several colleagues, students, and friends, both at Western and elsewhere. Among others, these include Tracey Adams, Anton Allahar, Robert Andersen, Doug Baer, Nora Bohnert, Sam Clark, George Comninel, Jim Conley, Jim Côté, Jim Curtis, Michael Gardiner, Ron Gillis, Neil Guppy, Monica Hwang, Bill Johnston, Fiona Kay, Julie McMullin, Kevin McQuillan, Jeff Reitz, Jim Rinehart, Jim Teevan, and Gloria Smith. Not all of these people will be fully satisfied with the final product, but the book has been made far better by their contributions to it. Writing can be a lonely endeavour, but any writing worth reading is almost always a collective enterprise.

Edward G. Grabb
London, Ontario, Canada
February 10, 2006

CHAPTER ONE

Theories of Social Inequality: An Introduction

It is a paradox of life that its familiar features are often the most difficult to comprehend. This poses a special problem for sociologists, who are expected to explain these familiar things but who are commonly accused of doing so in unduly complicated ways. And yet, while such accusations are sometimes justified, it is also true, as Randall Collins has noted, that obvious social questions may not have obvious or simple answers. Sociology's great strength, in fact, is precisely its potential for penetrating the superficial observation of everyday life and finding the fundamental social processes hidden beneath (Collins, 1992).

These remarks have a particular relevance for the topic of concern in this book—the problem of social inequality. For inequality is one of the most familiar facts of social life, a pervasive element in social relationships that is well known, even to the most casual observer. At the same time, however, social inequality has not been an easy problem to solve or explain. Instead, it has provoked protracted and perplexing arguments over whether it is good or bad, natural or contrived, permanent or transitory in social settings. Inequality has involved many of the most prominent social thinkers in debates about its origins, causes, and what, if anything, should be done to eliminate it. The central purpose of this volume is to review and assess the major elements in these debates, both past and present. In carrying out this purpose, the analysis will highlight the important differences and similarities in the approaches that have been developed, in an effort to trace the main direction that these theoretical exchanges seem to be taking us in understanding the problem of social inequality.

But what is social inequality, exactly? Even this most basic question is unlikely to produce a universally acceptable answer, although most would agree it involves such concerns as the gap between the rich and the poor, or the advantaged and the disadvantaged, in society. More generally, however, social inequality can refer to any of the differences among people (or the socially defined positions they occupy) that are *consequential* for the lives they lead, most particularly for the rights or opportunities they exercise and the rewards or privileges they enjoy. Of greatest importance here are those consequential differences that become *structured,* in the social sense of the term, that are built into the ways that people interact with one another on a recurring basis. Thus, people differ from one another in an infinite number of respects—in everything from height or weight to eye colour or shoe size—but only a finite subset of these differences will be consequential for establishing unequal relations among people, and fewer still will generate structured patterns that are more or less sustained over time and place.

This leads us to a second basic question: what are the main bases for inequality in society, the key differences among people that affect their rights, opportunities, rewards, and privileges? Once again, a universal answer is far from clear. Complications arise around this question for a variety of reasons.

First of all, some observers argue that the central bases for inequality are *individual differences* in natural abilities, motivation, willingness to work hard, and so on. In contrast, other analysts contend that inequality is primarily based on the differential treatment people are accorded because of *socially defined characteristics,* such as economic class, race, age, ethnicity, gender, or religion.

A second difficulty in trying to list the main bases for social inequality is the great discrepancy in the importance attached to some criteria across different places and historical periods. For example, religion historically was a major determinant of people's rights and rewards in countries like Canada or the United States but now is probably less significant than it once was. In other parts of the world, however, such as the Middle East or Northern Ireland, religious affiliation is still a key factor affecting differences in the economic well-being and political influence of people and continues to provoke grave conflicts between factions.

A third complication to consider is the disagreement that exists over how many bases for inequality are truly important, and which of these are the most consequential in any general portrayal of inequality in society. Opinions on this issue vary; some theorists see a whole spectrum of cross-cutting—individual and socially defined—inequalities, while others stress only one or two key bases as predominant over the rest. Marxist theorists, for example, often emphasize class inequality as the most important above all others, while some writers in feminist theory place much more stress on questions of gender inequality than on any other issue. (For discussion, see Rothman, 1993; Hurst, 1995; Curtis, Grabb, and Guppy, 2004; McMullin, 2004; Wright, 2005.)

Finally, the issue of what bases for inequality are most crucial in society is clouded further by the related issue of why some human differences are viewed as significant but not others. As an extreme case, we might wonder, for instance, why eye colour has virtually no impact in creating unequal social relations when something almost as superficial, colour of skin, has been used throughout history as a basis for granting or denying rights and rewards, thereby establishing persistent inequalities among people.

This list of issues provides us with some indication of the challenges involved in attempting to deal with the questions of what social inequality is about and where its major origins and bases should be traced. In subsequent chapters, we shall examine the ways in which leading social theorists have addressed such questions. Although a variety of themes and ideas will be considered, it is helpful at the beginning to be alert to certain topics that will be of particular concern. Hence, in the rest of this opening chapter, we shall briefly introduce four topics around which much of the later discussion revolves. These include the concept of class and its significance for the analysis of social inequality, the part played by the concept of power in theories of inequality, the role of the state in discussions of inequality in modern times, and the outlook for reducing or ending social inequalities in future societies.

CLASS AND SOCIAL INEQUALITY

While controversies abound in much of the literature on social inequality, one point that most observers accept is the central importance of the concept of *class* for the original development of theory and research in this area. Some have suggested that sociology itself, not just the analysis of inequality, really began with studies of class, especially in the work of Karl Marx (Dahrendorf, 1969; Giddens, 1973; Hunter, 1986). The term class, of course, is now part of the everyday vocabulary of many people. Research has shown that there is even some vague agreement in much of the population on how classes are perceived: as social groupings that differ mainly in their command of *economic* or *material resources,* such as money, wealth, or property (Coleman and Rainwater, 1978; Bell and Robinson, 1980; Grabb and Lambert, 1982; Lambert et al., 1986; see also Rothman, 1993: 106; Kelley and Evans, 1993).

Thus, it is no surprise that class has been central to most discussions of social inequality, since it pertains to that most basic of life's inequalities: the differential access of people to the material means of existence (Carchedi, 1987: 79). However, although most social thinkers can agree on the importance of the concept, numerous disputes and questions persist over the more precise meaning of class, how it should best be defined and what its real significance is in the general analysis of inequality. Among the many issues that have been raised, the following should be kept in mind in our review of the major theorists.

First of all, are classes simply *categories* of people in similar economic circumstances, or should the concept of class be reserved solely for those situations in which such categories are also real *groups,* with a common consciousness of membership and some basis for sustained interaction among the people involved? A somewhat related question is whether classes are equivalent to *strata,* to statistical aggregates of people ranked according to criteria such as income, or whether such delineations miss the essential significance of the term class (Stolzman and Gamberg, 1974).

Another issue of this sort concerns the distinction between *relational* and *distributive* studies of inequality: are classes most important or interesting for the uneven distribution of income and other rewards among them, or for the relationships of control and subordination that are established in their interactions with one another? (See, e.g., Goldthorpe, 1972; Grabb, 2004.) Fourth, are classes best understood concretely, as sets of real *people,* or abstractly, as sets of economic *places* or *positions,* filled by people and yet distinct from them, in much the same way that boxes are distinct from their contents? And finally, of course, there is the long-standing dispute over the number of classes that exist in modern societies. Are there just two, a small dominant one and a large subordinate one? Or are there instead some intermediate classes between top and bottom, and if so, how many? Or is it more accurate to say that there are *no* discernible classes, just a finely graded hierarchy without clear breaks? One of the purposes of this analysis is to sort out, if not completely resolve, the debates over these divergent views of class.

POWER AND SOCIAL INEQUALITY

Along with class, *power* has emerged as the other key concept in most major theories of social inequality. Like class, power is recognized as an important idea but at the same time has generated considerable debate over its exact meaning or significance. Largely through the early work of Max Weber, many theorists accept that power occurs where some people are able to control social situations as they see fit, whether other people accept or oppose this control. But many questions continue to arise around the concept. Can power differences between people be dispensed with under the appropriate social conditions, or are relations of control and subordination inherently necessary in any society? Is power unjust and oppressive by definition, or can it be a means for preventing or reducing injustice and oppression? Does power always flow downward from those who command to those who obey, or is there some reciprocal influence back up the chain of command? Such questions of power, of the ability to control social situations, obviously bear on the problem of social inequality, for such control normally has a direct effect in generating and sustaining unequal rights, opportunities, rewards, and privileges among people.

Yet, of all the important questions that have been asked about power, perhaps the most crucial one in recent years has been the following: from what and how many sources does power stem? In other words, does power ultimately originate in a single source or collectivity? Does it lie instead with a few principal factions or interest groups? Or is it widely dispersed to a multitude of individuals and groups in society? While there is still a range of opinion on this question among social theorists, many now seem to believe that power is considerably more concentrated than was once thought. Within this narrower range of viewpoints, one of the central debates is between those, on the one hand, who see power as primarily (even completely) a consequence of class differences and those, on the other hand, who see class as one basis (or the key basis) among a limited but nonetheless multiple set of power differences in society.

As shall be argued later, it is misleading to overstate the disagreements between these two competing viewpoints, since they share more elements in common than is often acknowledged. Still, there are some fundamental distinctions between them that have crucial implications for how we are to think about or conceive of inequality in society. The key difference is mainly in their emphases. The first view contends that social inequality is principally about class inequality and, though it acknowledges nonclass inequalities between races or between genders, for example, treats other inequalities as secondary to and largely explicable by class differences. The alternative view contends that while inequality is certainly about class, in some cases first and foremost, it is also about other socially defined characteristics such as race and gender. These traits and others are regularly tied to inequalities in power that are at least as significant as class inequalities in many instances, that are not reducible to class differences, and that operate whether or not classes exist in society. It is on this basis that some theorists see power as a more general concept than class, an idea that can be used, at a general level of analysis, to help us understand and explain social inequality in all its forms. The various disputes, as well as points of agreement, among leading class theorists and prominent power theorists form a major portion of the topics we shall address in this book.

THE STATE AND SOCIAL INEQUALITY

The choice of placing primary emphasis on class or on power when analyzing social inequality is roughly paralleled by a second debate that has taken on great significance in some discussions. This is the question of whether the economic structure is the key mechanism through which inequalities develop and are sustained in society, or whether instead it is the political structure, or *state*, that is the central arena for determining the nature of social inequality in modern times.

On the one hand, the economic structure, the system of material production, is the main engine for creating the necessities (and the luxuries) of social life; because of this essential activity, and because control over what is produced and how it is distributed is so closely tied to class differences, there is a case to be made for the view that the economic system and the class relations emerging from it form the crux of any analysis of social inequality. On the other hand, the state is supposed to be the official representative of the general will, at least in nominally democratic societies; hence, it has the responsibility for creating and implementing, by force if necessary, those laws and policies that can either entrench or reduce inequalities in society. This right to legislate and enforce gives to the state leadership ultimate *formal* control, not only over class inequality but over racial, sexual, and other forms of inequality, as well. Moreover, the pattern of extensive political intervention in economic affairs in many contemporary societies has prompted some observers to suggest that the state may ultimately usurp the essential activities of the economic structure. Such arguments provide the main basis for the view that the state has become the principal player in the modern power struggle and thus the predominant factor to consider in any analysis of social inequality.

THE FUTURE OF SOCIAL INEQUALITY

Probably the main reason why social theorists have devoted so much effort to the problem of social inequality has been their desire to answer one fundamental question: is inequality among people a natural and inevitable feature of society, something that might be reduced but never abolished in social settings, or, on the contrary, is inequality an unnatural and imposed social arrangement, an injustice that can and should be eliminated through social change?

The extensive discussions that have developed around this difficult question have given rise to several important controversies. Some theorists argue, for example, that inequality must be natural, since it is found in all societies. Others counter that its universal existence is no proof that inequality is natural; after all, slavery too was once a universal phenomenon but is now seen as an unnatural social arrangement that has been officially abolished by virtually all the nations of the world. A somewhat related issue is whether people should receive unequal rewards because they contribute different degrees of talent or effort to society, or whether it is possible to develop a social system where such inequality is essentially unnecessary, since people will have subordinated these self-interested motivations to a greater concern with serving the collective good. Still another question along these lines is the problem of the organizational necessity of social inequality: is it possible or not, in any complex society, to establish differences in tasks and administrative responsibilities that do not also promote persistent inequalities in the rights, opportunities, rewards, and privileges of people?

These are just some of the more prominent issues that underlie the general problem of social inequality. Although this introduction is not meant to be exhaustive, it should offer the reader an initial sense of the major topics and questions that theorists in this field have addressed. The points noted here will serve to guide and to structure most of the review and analysis that follow.

We begin in Chapter 2 with the work of Karl Marx, whose original thesis on class structure in capitalist society provides a fundamental backdrop for all of the other theories to be examined. In Chapter 3, Max Weber's attempt at a constructive critique and extension of Marx's views is the main concern. In Chapter 4, the work of Émile Durkheim is examined, and his role in the transition from Marx and Weber to the so-called structural-functionalist school is considered. In Chapter 5, the ideas that structural functionalists have put forth about inequality are assessed.

Chapter 6, which focuses on more recent theories, is the most diverse chapter in the book. Many of the theories considered in this chapter involve a critical reaction to, or rejection of, structural functionalism and a resurgence of interest in building from the ideas of Marx and Weber. Six writers, in particular, have been singled out in this regard: Ralf Dahrendorf, Gerhard Lenski, Nicos Poulantzas, Erik Olin Wright, Frank Parkin, and Anthony Giddens. In addition, Chapter 6 also provides a selective review of leading examples of theories that concentrate on two specific types of social inequality: gender and race. This special focus reflects the relatively large amount of attention that has been given to the concepts of gender and race by many contemporary theorists. One of the goals of this chapter will be to indicate some of the important areas of compatibility and congruence between leading examples of these specialized theories and the more general perspectives reviewed in the present analysis. In the last section of Chapter 6, we consider the question of transnational or global inequality, and discuss how inequalities among nations can also be understood by applying a general power perspective.

Finally, Chapter 7 provides an evaluation and assessment of the major elements in both the classical and contemporary perspectives. The volume concludes with some speculative observations on the contributions of existing theories to a more unified overview of social inequality, both in advanced societies and in the more general global context.

CHAPTER TWO

Karl Marx and the Theory of Class

> *Follow your own course and let people talk.*
> KARL MARX
> Introduction to Volume 1 of *Capital,* 1867

INTRODUCTION

We begin our consideration of the major theories of social inequality with the work of Karl Marx. This choice may seem inappropriate to those who suggest that Marx was not a theorist of general social inequality at all (Benson, 1978: 1). Thus, for example, Marx showed no sustained interest in examining the numerous ways in which people are ranked in societies at any one time. Nevertheless, there is no doubt that, for several reasons, an understanding of Marx's work is a crucial basis for developing an overall analysis of social inequality in advanced societies.

First of all, Marx has been characterized by various analysts as the father of modern sociology itself (e.g., Berlin, 1963: 158; Singer, 1980: 1), although Auguste Comte usually gets credit for this role and others, most notably Saint-Simon, may have an equal or stronger claim (Durkheim, 1896: 104; Giddens, 1973: 23). Because of Marx's central position in the general development of sociology, any of his work that is relevant to the study of social inequality necessarily should be examined and assessed.

A second reason why Marx's work is essential is that he, more than any other thinker, is responsible for bringing to the forefront of sociology the concept of class. Some claim that this concept has become obsolete or outdated in social theory (Nisbet, 1959; Clark and Lipset, 1991; Clark, Lipset, and Rempel, 1993; Pakulski, 2005). Others argue, on the contrary, that class continues to be one of the principal explanatory variables to consider when trying to understand the workings of society (e.g., Giddens, 1973: 19; Hout, Brooks, and Manza, 1993; Wright, 1994: 19; 1997: 1–3; 2005: chap. 1). In either event, it is true that the idea of class has had a significant influence on the way that people think about society and their own positions in it. Academics routinely employ the term in their research on the operation of social structures, while people regularly allude to class or related ideas in their everyday conversation. It is the use, and frequent misuse, of the concept of class that forms much of what until recent times has constituted the study of social inequality. Thus, Marx's original formulation of the theory of classes and class structure is a critical starting point for our discussion.

A third reason for Marx's importance is that his analysis of class structure, coupled with his conclusion that future societies will see an end to class-based inequality, has helped to stimulate, or provoke, virtually all of the attempts by subsequent social thinkers to explain social inequality and its transformation over time. Marx shared with later thinkers such as Weber and Durkheim a keen interest in the process by which earlier forms of society, particularly feudalism, eventually gave rise to the contemporary form of society known as advanced capitalism. Marx's view was that the capitalist stage of social development would itself ultimately be superseded by the final and highest form of society, alternately labelled *socialism* or *communism*.

It is on the questions of why societies change, and what their ultimate form will be, that many later theorists seem to take issue with Marx. In fact, what has been called "the debate with Marx's ghost" (Salomon, 1945: 596; Zeitlin, 1968: 109; Giddens, 1971: 185; Sayer, 1991: 13) frequently amounts to a debate with certain Marxists— individuals who have adopted Marx's ideas as their own and have interpreted or modified his original work in a variety of ways. As is the case with other schools of social thought, too often these many versions of Marxism have proven internally inconsistent, even contradictory, leading Marx himself to utter his famous remark: "What is certain is that I am no Marxist" (quoted by Engels, 1882: 388). The confusion surrounding what Marx really said, or at least really meant, inevitably makes the understanding of his work more difficult. Nevertheless, it will be argued here that the central features of Marx's theory, particularly as they pertain to the study of social inequality, are not as obscure as some analysts suggest. Moreover, the wide range of reactions both for and against Marx's work says as much about the power of his ideas as it does about their complexity.

Until the last decade of the twentieth century, nearly four of every ten people in the world lived under governments that at least claimed to follow Marxian principles. This fact is perhaps the most telling evidence of his influence. Even now, in spite of the decline of Marxist-inspired governments throughout most of the world, there can be little doubt that Marx's ideas retain much of their power to stimulate debate and inform discussion on the nature of human societies (e.g., Lemke and Marks, 1992; Boswell and Dixon, 1993). Thus, while Marxism is undoubtedly in decline as a political force in the current era, claims for the imminent death of Marxism as an intellectual force seem clearly to be exaggerated (see, e.g., Wright, 1994: chaps. 8 and 11; 2005: chap. 1; Wallerstein, 1995, 1998).

Of course, Marx himself did not live to see the birth of any society ruled by the socialist or communist precepts that he championed. Marx was a nineteenth-century man who devoted his intellectual efforts toward understanding the workings of capitalist society at that time, with the hope that this advanced but flawed form of social organization could be radically transformed to suit the needs of all people. His special concern was European, especially English, capitalism in the 1800s. Hence, some have alleged that his analysis is not totally applicable to present-day societies, societies that have been variously labelled "monopoly capitalist," "state capitalist," or "postcapitalist" (e.g., Dahrendorf, 1959; Baran and Sweezy, 1966). Even so, although the capitalism Marx knew may in some respects no longer exist, in other respects it seems, like the poor and the exploited, to be very much with us still. Capitalism, in its many current guises, continues to be the pre-dominant mode of economic and social organization in the world. The pre-eminence of the capitalist system seems even more evident now than in the past, given the demise of so many socialist regimes in recent years. Nevertheless, whether Marx's vision is truly dead or, instead, is simply awaiting its resurrection, remains an open question. In the meantime, the disagreements between proponents of capitalism and socialism still inspire many of the major contemporary debates surrounding the problem of social inequality. Marx is such an important figure precisely because his analysis of capitalism, as well as his prognosis that socialism is what the future holds for all of us, situates his work at the centre of this fundamental controversy.

For all these reasons, then, Marx's theory of class and class structure in capitalist society is our first concern in this investigation. In order to grasp more fully the origins of Marx's principal ideas, it is instructive to begin with a brief examination of some of the major events and intel-lectual influences in his life.

BIOGRAPHICAL AND INTELLECTUAL SKETCH

Karl Marx was born in Trier, Prussia, in the German Rhineland, in 1818. He died in London, England, in 1883. These 65 years were turbulent ones in Europe, marked by considerable social change, economic development, political upheaval, and intellectual debate. It is not possible here to trace all the numerous people and events that helped shape Marx's ideas; nevertheless, several major influences should be identified for their per-sistent and deep-seated impact on his thought.

Early Years

Perhaps the first major influence on Marx's ideas occurred during his adolescence. Marx's father was a moderately prosperous lawyer of Jewish background, well educated and a believer in the liberal ideals of the French Revolution of 1789—individual freedom and equality in particular. Within his father's circle of intellectual middle-class friends, Marx was exposed to these egalitarian views of people and society. As a result, Marx developed a particularly optimistic philosophy about the nature of humanity and the potential of civilization to progress. Later in life, his father would reverse himself and embrace the conservative and reactionary ideas of the Prussian government under which he lived. Part of this reversal included renouncing Judaism for Protestantism, as a means of ingratiating himself with the Prussian authorities. However, throughout his life, Marx himself retained the belief, learned in these early years, that people are by nature both good and rational. Human beings have, to Marx, the potential to produce a nearly perfect society if only unnatural obstacles, especially exploitation and oppression by those in economic and political control, are removed from their path (Berlin, 1963: 29; McLellan, 1973: 87; Singer, 1980: 22). As we shall see, Marx's belief in human progress and perfectibility had a significant impact on his expectations for the transformation of the capitalist society in which he lived and for the socialist society of the future that he envisioned.

University Experiences

The second important intellectual influence on Marx came into play during his university education. Initially Marx enrolled as a law student at the University of Bonn in 1835. Accounts suggest that the seventeen-year-old Marx started out living the "dissipated life of the ordinary German student"—drinking, writing poetry, getting slightly wounded in a duel, and being arrested occasionally for "riotous behavior" (Berlin, 1963: 33; Singer, 1980: 2). After a year of this, Marx's father convinced him to transfer to the University of Berlin; however, this radically changed the course of Marx's life and ended his father's hopes that Marx would become a lawyer.

At Berlin, Marx became absorbed in the ideas of the philosopher Georg Hegel. Hegel had died in 1831, but his thought was still a dominant force in the intellectual life of German universities at the time. This was particularly true at Berlin, where Hegel had taught for many years. Marx soon switched from law to the study of Hegelian philosophy. He eventually rejected the content of Hegel's theories, because of their preoccupation with the abstract world of the mind or the spirit and their tendency to treat ideas as mere mental constructs divorced from the real, concrete world of people and society. Nevertheless, Marx was to the end of his life strongly influenced by the *form* or *method* of Hegel's thought. First of all, he accepted completely Hegel's emphasis on the importance of understanding the past, of studying history in order to comprehend the present and project the future (Giddens, 1971: 4). Second, Marx employed for his own more concrete purposes the famous and often misunderstood Hegelian notion that change occurs in a *dialectical* fashion. Not all writers agree on the meaning of this idea (Elster, 1985: 44). However, put simply, to say that change is dialectical indicates that it is by nature a process of struggle or tension between opposing, incompatible forces. Moreover, in the clash of these forces, neither side comes out in the end unchanged or unscathed. When the struggle is resolved, the resulting victorious force is, in a sense, different from either of the original adversaries, more highly developed and advanced than each predecessor. This dynamic conflict of forces is continual, with newly emerging forces eventually entering the arena to battle the latest victor (Berlin, 1963: 55–56).

The problem with this notion, in Marx's view, is that Hegel makes the error of applying it only in the airy realm of ideas—a fact that is consistent with Hegel's tendency to disregard material, concrete phenomena. Marx's great task at this stage in his life became to "turn Hegel on his head," to take this dialectical method for thinking about change out of the clouds and to apply it to the understanding of how real societies and real human groups, arise, develop, and become transformed over time. Hence, for Marx the key opposing forces to examine are groups, not ideas. Ideas are important social forces, no doubt, but cannot be divorced from the social context in which they emerge or from the people who think them. Marx's application of Hegel's historical and dialectical view of change will become evident later when we consider Marx's analysis of class and class struggle in capitalist society.

Political Activist and Refugee

Marx completed a doctoral degree in philosophy in 1841. He ultimately received his degree, not from the University of Berlin but from the smaller University of Jena, where he expected there would be less opposition or delay in the acceptance of his doctoral thesis (McLellan, 1973: 40). Marx then hoped to obtain a university teaching post; however, because his political views were already known and unpopular with the conservative Prussian government, the ministry of education ensured that no teaching position was forthcoming. Marx then found alternative employment as a journalist, an occupation that he held on several occasions throughout his life. His first job of this type, like virtually all the others, was short-lived. As writer and eventual editor of the *Rhenish Gazette* (*Rheinische Zeitung*), he published articles that criticized the Prussian government, especially for its treatment of the poor and working classes. This criticism led to the censorship of the paper in 1843. By then, Marx had become convinced that the same working class he was writing about was destined to be the prime mover in the progress toward a new universal society, where true freedom and full human potential would be realized for all. He now saw himself as playing a key role in awakening the largely unaware labouring masses to their revolutionary task.

In order to express his ideas and fulfill this role more freely, Marx left Germany for France. Initially he attempted to publish a joint French-German journal in Paris, but it met with indifference in France and quick suppression in Germany. Marx lived the years between 1843 and 1849 as a political refugee, first in France, then in Belgium, and briefly in Germany. During this period, Marx was exposed to several crucial influences on his thought. Out of Ludwig Feuerbach's critique of religion as an alienating force for humanity, Marx developed the concept of alienation, and, as we shall see, he eventually applied this idea to his analysis of the plight of workers in capitalist society. Marx also read the works of several French socialists. These confirmed in his mind the important role of the working class in transforming capitalism, even though he disagreed with most of these writers on specific issues.

Probably the strongest influence on Marx in this period was Claude-Henri de Rouvroy, Comte de Saint-Simon, a French social thinker of the late eighteenth and early nineteenth centuries whose views on social change corresponded closely with those evolving in Marx's own mind. It was largely from Saint-Simon that Marx came to see economic relationships as the key to explaining historical social change. Marx's emphasis on the role of conflict, especially between economic classes, is also in part adopted from Saint-Simon, as is the view that social change occurs because each society bears within it, and generates through its own development, the "germ of its own destruction" (Giddens, 1973: 23; Berlin, 1963: 90–91). The parallels between the latter idea and Hegel's dialectical view of change were apparent to Marx and served to solidify his views on this issue.

The most exciting political activities and events of Marx's life also occurred in this turbulent period. By this time, Marx had met his lifelong friend and collaborator, Friedrich Engels, and together they composed a short pamphlet laying out the doctrines of the Communist League, newly formed in 1847. This work was published in 1848 as the now famous *Communist Manifesto*. In that same year, popular revolts against established monarchies occurred in parts of France, Italy, Germany, Austria, and Hungary. Much of Europe was astir with the desire for liberal reform, if not radical change. However, none of these uprisings was completely successful, and any hopes that Marx and others may have had that these events were the forerunners of a socialist revolution were soon quashed. By the middle of 1849, Marx was again on the move, this time to London, England. Despite occasional glimmers of political upheaval on the

Karl Marx, 1818–1883

Portrait of Karl Marx (1818–1883)
(b/w photo), NO_DATA/Private Collection/
Bridgeman Art Library.

continent, the revolution that Marx awaited did not take place, and he was to remain in England for the rest of his life.

England and *Capital*

In the more than 30 years during which Marx lived in England, his principal achievements were those of a scholar. His activities as a radical journalist, political agitator, and organizer of the working class were greatly reduced, although he did take a leading hand in the operation of the first International Workingmen's Association between 1864 and 1872. What consumed most of Marx's energy was his monumental analysis of contemporary capitalism.

Marx, in fact, had contracted to do such a work as early as 1845, but it was repeatedly set aside for his political activities and other writing (Singer, 1980: 4). It was both appropriate and somewhat ironic that Marx found himself in England during the writing of this work. England was the most advanced capitalist society in the world at the time and hence served as the most appropriate example or archetype for Marx's analysis. It was ironic, however, that Marx, a radical opponent of capitalism, should find tolerance and refuge within which to express his ideas only in England, the prime example of what he most despised (Berlin, 1963: 181).

What Marx found particularly disturbing about capitalism, both in England and elsewhere, was that so much surplus wealth could be produced by this system and yet be distributed to such a small group of people. While the capitalists, the owners of large factories and agricultural lands, were themselves living in splendour and affluence, the workers, who toiled to produce this great wealth, subsisted for the most part in miserable poverty. The contrast between rich and poor was particularly acute in Marx's time, with labourers sometimes working eighteen hours daily. Six-day work weeks were typical, and child labour—employment of children as young as six or seven—was not uncommon. Working conditions themselves were often of the most appalling kind, dirty, cramped, and extremely hazardous (see Marx, 1867: chaps. 10 and 15).

Marx's condemnation of the capitalist system in his work *Capital* (*Das Kapital*) was based on the statistical and other evidence made available to him during his long days of study in the British Museum, and on his own observations and first-hand experience with poverty. Marx and his family at times endured economic difficulties during this final stage of his life, although some inheritances and the help of Engels and other friends lightened the burden.

Marx's intermittent economic plight was one factor that slowed his progress on the analysis of the capitalist system in England. His emotional strength and resolve to finish the work were also sapped by illness and personal tragedy, including the premature death of his wife and several of his children. Thus, while *Capital* was intended to be the crowning work of his life, only the first volume was actually completed, with the second and third volumes compiled and edited by Engels from rough drafts and notes. It would appear that Marx was on the verge of outlining explicitly his most crucial concern, a detailed treatment of the class structure in capitalism, when he died in his study on March 14, 1883.

The incomplete nature of Marx's analysis of capitalism is indeed unfortunate and has only added to the uncertainty and controversy surrounding his work. As noted earlier, however, it is possible to piece together from the existing material, including the *Communist Manifesto* and his numerous other writings, a more or less consistent and comprehensible theory of class, class structure, and social change. A discussion of this theory is our main concern in the remainder of this chapter.

MARX'S THEORY: CLASS, CLASS STRUGGLE, AND HISTORICAL CHANGE

"The history of all hitherto existing society is the history of class struggles" (Marx and Engels, 1848: 57). In this straightforward statement at the beginning of the *Communist Manifesto,* Marx provides a key to understanding the essence of his theory. While a massive and complex literature has been generated by both critics and disciples in reaction to

Marx's work, this single assertion captures what may be the most crucial element in his conception of social structure and social change.

In Marx's view, historical epochs are distinguishable from one another primarily by the system of economic organization, or *mode of production,* that dominates in each era. Thus, the slave-based economies of ancient Greece and Rome are different in key respects from the feudalism of medieval Europe and the capitalist system of Marx's own time. What is similar about all these historical examples of social and economic systems, however, is that according to Marx each one is marked by a basic distinction between classes, between groupings of people who differ in the roles they play in the productive system. In all cases, the most important difference is between those who have *property* and those who do not. Property involves the right of ownership. However, in this context, property refers not to simple personal possessions but to resources that can be used to produce things of value and to generate wealth: land, rental housing, machinery, factories, and the like (Marx and Engels, 1846: 230).

The basic split between the owning, propertied class and the non-owning, propertyless class is always present, though it has taken different forms historically: masters versus slaves in ancient times, lords versus serfs in feudalism, the *bourgeoisie* versus the *proletariat* in capitalism. Certainly, more detailed divisions and distinctions might be drawn in each form; nevertheless, the two-class model is for Marx the crucial one to apply in understanding how societies are structured and how they change (Marx and Engels, 1848: 58).

Why is the two-class view the key to understanding society? Primarily because, according to Marx, all societies are born of struggle, of the underlying tension or open conflict between the two major economic classes. In fact, social change or development historically has never occurred without it: "No antagonism, no progress. This is the law that civilization has followed up to our days" (Marx, 1847: 132). To Marx, owners and nonowners have interests that are naturally opposed to one another. Whenever there is a situation in which one group controls the property, the productive apparatus of society, the remaining population is by definition excluded from such control. Nonowners necessarily enter into an exchange relationship with the owning class, giving their labour power to the owners in return for sufficient income to survive and to continue working in the employ of the owning class. The coercive and exploitative aspects of such a system are obvious in the slavery of ancient times and the serfdom of feudalism, since slaves and serfs were legally bound to labour for their masters or manor lords and had virtually no power to alter this arrangement. Coercion and exploitation may seem absent in capitalism, since the exchange of labour for wages between workers and owners occurs between people who are legally free to choose whether or not to accept such a contract. For Marx, however, the propertyless worker in capitalism is no more at liberty to withhold his or her labour from the owner than was the slave or the serf, because the alternative to wage labour is starvation if one possesses no productive property with which to provide food, clothing, and shelter–the simple means of survival.

It is apparent, then, that Marx's theory portrays all societies up to and including his own as arenas for an elemental struggle between "haves" and "have nots," between "oppressor and oppressed" (Marx and Engels, 1848: 58, 92). Societies are based on this inherent dichotomy in the productive sphere, and the major force for change in society necessarily involves a conflict of interests between these two groups. We will examine more closely just how Marx envisions the role of conflict in social change. Before doing this, however, we should consider Marx's views on the work process, especially in the capitalist system that was his principal focus. The operation of the capitalist mode of production and its consequences for the people involved in it will be of special concern in the section that follows.

Work in Capitalist Society

In our discussion of Marx's background at the beginning of this chapter, we noted his belief in the potential that humans possess for building increasingly better forms of society. Marx sees the best evidence for this in humanity's increasing mastery over nature and the material world throughout history. Human improvement is especially apparent in our ability to produce the means of life for a growing population. Compared with simple tribal societies, for example, modern nations, with their advanced tools and technical knowledge, are far more capable of providing for the needs of people. In fact, modern societies are increasingly able to generate enough goods to accumulate a large surplus of wealth over and above basic needs.

To Marx, it is productive work, not leisure, that under the right conditions is the ideal human activity (Avineri, 1968: 104, 107). In fact, it is this creative capacity, this ability to produce things of value, that is essential to human nature and that mainly distinguishes humans from other life forms (Marx and Engels, 1846: 31; Marx, 1867: 177–178). The ability of modern societies, then, to accumulate a vast surplus of produced wealth is itself a grand accomplishment and a tribute to human potential. That capitalism is the system of economic organization that has been most successful in generating wealth is acknowledged and even applauded by Marx (Marx and Engels, 1848: 65–66). The problem with capitalism, however, is that it distorts the structure and meaning of the work process, with negative consequences for society as a whole and for workers in particular. The distortion comes from several characteristics basic to the capitalist mode of economic organization: private property, surplus expropriation, the division of labour, and the alienation of work.

Private Property

Economic production is by definition a social activity, requiring groups of people working together to create things. To Marx, the social character of work is wholly consistent with the general communal nature, or "communist essence," of people and belies any claim that each of us is an isolated individual in society (Marx, 1858: 84). But the capitalist mode of economic organization artificially creates a situation in which people are individualized and separated in their work, even though, paradoxically, they may be working side by side.

The first factor at the root of the fragmentation of work in capitalism is the existence of private property. As was mentioned earlier, private property is mainly responsible for creating the two-class system in capitalism, the distinction between the owners of property, or bourgeoisie, and those who work for the owners, the proletariat. Marx believes that this initial split immediately erodes the naturally social character of production, to the extent that those involved are not working as a united group toward some common goal but rather are pursuing their own special interests. The principal interest of the bourgeois owners in this context is to maximize their own wealth rather than the general wealth or that of their workers. However, the workers themselves also act in their own individual interests in capitalist society and not for the general good (Marx, 1844: 238).

In Marx's time, workers lacked bargaining power through unions or legal strikes. As a result, they could not form a united front against employers, and their interests were rarely served. Typically, the supply of available labourers exceeded the demand, especially with the development of mechanized production methods. Hence, employers had a ready supply of unemployed workers willing to take any position left vacant by any worker who was fired for making trouble or because of illness or injury. This idle "reserve army of labor," as Marx called it (Marx, 1867: 487, 632), ensured that each worker acted primarily as an isolated person, in competition with other workers for a job and unlikely to want to band together with others to seek better pay or working conditions. This isolation of workers from one another obviously worked to the advantage of the capitalists, since it assured them of a docile, though frequently discontented, pool of cheap labour.

Expropriation of Surplus Wealth

A second aspect of capitalism that Marx believed distorts the natural work process is the expropriation of surplus wealth by the capitalist at the expense of the worker. One of the great ironies and injustices of capitalism, in Marx's view, is that working people are, through their labour, primarily responsible for the wealth that is generated, yet workers receive only a small portion of this wealth. To Marx, the value that a produced object has is the value of the labour used in its creation. This value should rightly belong to the workers whose labour is expended in making the object. But, under capitalism, workers exchange their labour power for wages that amount to less than the value of the object when it is sold in the marketplace. The difference between the wages and the value of the object when it is sold represents a surplus amount that goes to the capitalists as a profit, after their expenses are deducted. The capitalists treat this excess as their own and think of the wages paid out to workers as one of the costs of production.

To Marx, this perception of the process is completely incorrect. It is the workers who have incurred a cost in the exchange of labour for wages, not the capitalists. The workers lose part of the value of their own labour in the surplus that the capitalists usurp. This surplus, which is often reinvested by the capitalists, is in fact the accumulated "dead labor" of past workers. Capital is really this accumulated past labour, which, in the purchase of new machinery and so on, is used by the capitalist to expand control over and exploitation of "living labor"—that is, the current crop of employed workers (Marx and Engels, 1848: 84).

The Division of Labour

To this point we have discussed two characteristics of capitalism that according to Marx have distorted the essentially social nature of people and of work: the role of private property in fragmenting both capitalists and workers into self-interested individuals, and the unjust expropriation of wealth by capitalists at the expense of workers. A third feature of the capitalist mode of production, and one that intensifies these problems still further, is the division of labour.

In its simplest form, the division of labour arises as soon as private property is instituted and the class of owners is separated from the class of workers. Immediately there is a splitting up of those who own the means of production and oversee its operation, on the one hand, and those who labour within the production process, on the other hand. As capitalism evolves and develops as a mode of economic organization, this simple division becomes more and more elaborate. It is increasingly apparent that goods can be produced more quickly and efficiently if all the tasks that go into creating an object are divided into specific jobs, each of which is done by a particular worker on a continual basis.

The modern automobile assembly line is perhaps the epitome of how capitalism subdivides the production of an object into a series of small tasks. Capitalist production, based on a complex division of labour, is highly efficient, with far fewer hours of labour required to create the same volume of goods. However, this efficiency benefits the few capitalists at the expense of the many workers, in Marx's view. Workers produce more in the same working time but are paid the same wages. In effect, then, workers have exchanged their labour power for a smaller proportion of the wealth produced. The surplus wealth generated is larger and hence so are the profits of the capitalist.

Not only does the division of labour increase the profits of the capitalist, and thus the rate of exploitation of the workers by the capitalist, but the division of labour also has severe negative effects on the workers' whole orientation to work. We have already noted that Marx believed humanity's essential ability resides in creative labour. Work is an inherently enjoyable activity under natural conditions, since it is the means by which human beings create and shape their own world. But the division of labour into a series of repetitive and routine tasks is destructive of this enjoyable quality of labour. The pleasure of making things is lost to the workers, who find it virtually impossible to identify their productive power in an object to which so little of the creative self has been applied. Hence, for example, putting the same bolts on a series of automobiles all day, every day, provides none of the pride in workmanship and sense of creating something that go into the making of a custom car by a few people.

Alienated Labour

Thus, in the operation of the capitalist mode of production, work becomes something grotesque. Each person's special capability as a creating and producing being is turned into something that divides and separates people from one another, that is used by some people to exploit others, and that, as an activity, is hateful drudgery to be avoided whenever possible.

Marx uses the concept of *alienation,* or *alienated labour,* to signify these negative aspects of work in capitalist society. Alienation, unfortunately, is a widely misused term, employed by some analysts to label various types of psychological malaise or personal dissatisfaction. In fact, what Marx means by alienation is something quite different. At its most general level, alienation denotes the separation of something from something else (Schacht, 1970). Whenever something that is an attribute or a creation of people is somehow taken away from them or made external to them, in a sense we have a situation of alienation. Marx first encountered this idea in his early studies of Hegel. In particular, Marx was influenced by a critique of Hegel's religious views by the philosopher Ludwig Feuerbach. Religion, to Feuerbach, involves the attribution of special qualities and powers to a supreme being or god: mercy, compassion, knowledge, the power to create, and so on. In fact, Feuerbach claims, these qualities are humanity's own highest traits and powers, falsely separated or alienated from humans and projected onto this mythical god (Giddens, 1971: 3–4).

Marx agrees with Feuerbach on this point, asserting that, in religion, "the more man puts into God, the less he retains in himself" (Marx, 1844: 272).

Marx also sees a comparable process of alienation in the field of human labour under capitalism. We have so far discussed three features of capitalist economic organization: the existence of private property, the expropriation of surplus by the capitalist class, and the division of labour. All of these contribute in part to alienation. Marx's conception of alienated labour is evident from at least four key points in his analysis (Giddens, 1971: 12–13).

Alienation is apparent, first, in the separation of workers from control over the products they create. These products are the physical embodiment of human labour given by the workers to the capitalist in exchange for wages. This exchange means that labour has itself been alienated from the workers, taken from them in this sense, and then sold as a commodity that is basically no different from the commodities or goods that are created (Marx, 1844: 272; 1867: 170; Marx and Engels, 1848: 68). In addition, most of the wealth that these products can garner in the market is also taken away from the workers, expropriated by the capitalist and turned into profit.

A second feature of alienation involves the actual work task, which loses its intrinsically enjoyable and rewarding character in capitalism. The division of labour into specialized, repetitive tasks, coupled with the lack of control over the work process in such a setting, means that the worker "does not feel content, but unhappy" and "does not develop freely his mental and physical energy" (Marx, 1844: 274). Work does not satisfy the human need to create but becomes an alienated, external means to satisfy other needs. Leisure time and the "animal functions" of "eating, drinking, and procreating" become most important to alienated people, while work, the essential human activity, "is shunned like the plague" (Marx, 1844: 274–275).

A third aspect of capitalist production that promotes alienation is also related to the division of labour, especially to the increasing technological complexity that the division of labour entails. Machine technology, for example, is itself an important illustration of how people can harness natural forces and materials to perform useful tasks. Active human mastery of the world is apparent in this mechanization. Ironically, however, the requirement by capitalism that workers must operate this mechanized apparatus on a continual, routine, and repetitive basis means that people are effectively subordinated to the machines that humans have created (Marx, 1844: 273; 1867: 645; Marx and Engels, 1848: 69). Once again, then, people's labour takes on a separate, alienated form in the machines that dominate their own creators.

Finally, a fourth sense in which Marx identifies alienation in capitalism concerns money and its power literally to buy anything, even people and human relationships. Marx states: "I am ugly, but I can buy for myself the most beautiful of women, therefore, I am not ugly"—if, that is, I have the power that money brings (Marx, 1844: 324). The power of money is itself evidence of how human qualities, such as beauty in this example, are externalized from humans and converted into fluid, quantifiable assets like any other commodity. Money is "the alienated *ability of mankind*" (Marx, 1844: 325), set apart from people themselves. Here again we see the idea of externalization or separation that is basic to the concept of alienation.

Marx believes that alienation in all these forms is inherent in the capitalist mode of economic organization and can be ended only by the overthrow of the capitalist system. Marx is convinced that certain conditions necessary for such a revolutionary overthrow will emerge naturally from the internal logic of capitalist development. In the next section, the major points Marx raises in this regard will be considered.

Capitalist Development and Change

We have already discussed the importance that Marx places on the struggle between opposing classes throughout history. This emphasis on the conflict between opposites is consistent with the Hegelian dialectical perspective that influenced Marx's early thinking about change. Marx combines this dialectical method with a similar conception borrowed partly from Saint-Simon: the idea that every societal form holds within itself "the germ of its own destruction"—some force or group that eventually arises out of the system that currently exists, opposes the present ruling regime, and thereby changes society in some fundamental way. Hence, it is always the case that "within the old society the elements of a new one have been created" (Marx and Engels, 1848: 91).

Marx applies these conceptions in his analysis of how capitalism will develop and ultimately be transformed into socialism. Marx begins, however, by examining feudalism, the societal form that preceded capitalism in Europe. In his view, an understanding of how capitalism came to emerge from the ruins of feudalism helps to clarify just how socialist society can develop out of capitalism.

Feudalism and Capitalism

Feudal society in medieval Europe was a highly localized and community-oriented social structure, based on a rural, agricultural economy. Each serf was bonded to a manor lord and worked the lord's land in return for being permitted to live as a tenant on the lord's estate. This system was relatively stable for centuries but gradually evolved and ultimately disintegrated because of a combination of factors.

One factor was the voyages of discovery during this period, which opened up international trade, expanded markets beyond the local communities or feudal estates, and created demands for new goods. The new markets and new demands stimulated the development of new technology and more efficient production to meet these requirements (e.g., Marx and Engels, 1848: 63–65). The growth of large-scale manufacturing thus began and with it a demand for labour to work in these settings. Similar changes occurred in agriculture, with the move toward large-scale farms, more sophisticated production techniques, and the increasing use of wage labourers rather than serfs to work the land (e.g., Marx, 1867: 742; see Hopcroft, 1994).

Increasingly, the old feudal economy was unable to compete with this new mode of production, to hold onto traditional customers for its goods, or to acquire any share in the expanding markets for new goods. Feudal lords also saw their resources depleted by wars and civil strife. The impoverished aristocracy became less and less able to afford to retain serfs on their lands. Many serfs, therefore, were thrown off the land or else, intrigued by the broadening horizons of this new age, sought their own fortunes voluntarily as free people. Many of the freed serfs migrated to the growing urban centres, where they took up employment as workers for the emerging manufacturing interests located in these centres. Such workers were the beginnings of the proletariat, while the merchants, traders, and financiers involved in the manufacturing ventures were the basis for the capitalist bourgeoisie (e.g., Marx and Engels, 1846: 33–35, 65–70; 1848: 59–60; Marx, 1867: 745; see also Avineri, 1968: 151–156; Giddens, 1971: 29–34).

As a result of these historical and other developments, capitalism eventually emerged within the feudal society it was to replace. The transformation, of course, occurred over several centuries and did not happen everywhere at the same pace. Ultimately, however, it took place throughout Europe. To Marx, the opposition of interests between the developing class of capitalists and the existing feudal aristocracy became increasingly evident at this time. Marx viewed the capitalist class in this era is as a dynamic force for social and economic change, creating new ideas, new production methods, and new wealth. In contrast, the feudal lords are seen by Marx as increasingly superfluous to the productive system, even though they continued to reap wealth from it because of the political and other privileges they enjoyed under monarchical governments. In the end, though, the capitalist class threw off the feudal yoke that was hindering its progress. This change sometimes came about through a combination of confrontation and political accommodation, as in seventeenth-century England, and sometimes through violent revolution, as in eighteenth-century France (Marx and Engels, 1846: 72–73).

Modern observers now dispute several aspects of Marx's analysis. They point out, for example, that many of the new capitalists of this period were themselves members of the feudal aristocracy (Wallerstein,

1979: 142; Collins, 1980: 939; Goldstone, 1986: 258–267). There is also a suggestion that the key conflicts that promoted the transition from feudalism to capitalism were mainly between different elite groups within the feudal aristocracy, rather than between capitalists and feudal lords (Lachmann, 1990).

Nevertheless, such disputes do not contradict Marx's basic contention that capitalism emerged from conditions arising within feudalism, or that a similar process could also occur within capitalism. Marx believed that, in this same conflict-driven, dialectical fashion, capitalism will be transformed by a new opposing force growing within itself: the proletariat. His view is that the dynamic and progressive nature of the capitalist class—qualities that played such an important part in overthrowing feudalism—will gradually dissipate as capitalism matures as a system. The entrepreneurial spirit, speculative daring, technical knowledge, and administrative skill displayed by the capitalists will have served their purpose at this point. According to Marx, these qualities will be replaced by increasing entrenchment and resistance to change among the bourgeoisie, who, having taken power, will cling tenaciously to it. During the same period of bourgeois rule, the mass of the population, the working class, will become aware that they themselves, not the bourgeoisie, are the force primarily responsible for the wealth generated in capitalism.

Marx believes it is the destiny of the workers and the producers of society to take the ascendancy and thereby transform the entire economic and social structure. This change will signal the beginning of the final stage of human development: socialist or communist society. This will be a unique achievement in history, for it will be the first successful revolution in which the victors are the majority of the people and not some select minority. It will also mean an end to class divisions and class struggle, since only one class, the universal proletarian class, will remain (Marx and Engels, 1848: 77; see Avineri, 1968: 60–62).

The Factors in Capitalist Transformation

We have not yet specified those processes Marx perceives in capitalism that will promote its demise. Though there are different interpretations and numerous factors that could be noted, seven points in particular seem crucial (Giddens, 1971: 52–60; 1973: 34–37).

First of all, the rise of capitalism, as noted earlier, is closely tied to the growth of large urban centres, which develop as propertyless rural inhabitants flow to the cities in search of employment. This growing concentration of urban labourers is a key initial step toward the development of working-class awareness, since masses of workers are housed in the same neighbourhoods and experience similar living conditions.

A second, related process is the increasing expansion of production into large-scale factories employing many workers. This process makes apparent the basically *social* nature of human production. It also means a concentration of workers in close proximity to one another and a further basis for awareness of their common class position and their common plight (Marx and Engels, 1848: 69-70).

Marx suggests a third factor by which workers are made more aware of their shared fate and their separateness from the capitalist class. This factor is the greater suffering that workers experience during economic crises. Marx's view is that capitalism is primarily concerned with a rational and efficient, but relentless and self-interested, pursuit of *profits* by the owning class. The drive for more and more profits frequently produces unfortunate consequences for society as a whole, but especially for workers. The bourgeoisie's desire for greater gain often means that it produces more goods than it can sell. This imbalance leads to a loss of money by capitalists, since they must sell their products in order to realize any profit and pay their costs. A common reaction by the capitalist class is to lay off workers in order to reduce costs and save money. But the workers of society are also the main purchasers of goods. Hence, laying off workers means that there are fewer people able to buy products, so that there is an even lower demand for goods and a further loss of money for the capitalists. In this spiral fashion, economic crises tend to occur regularly in the capitalist system and in fact are basic to its operation. Overproduction, followed by higher unemployment, reduced demand, slow economic growth, and depression is the typical pattern. After each crisis, there is a period in which the economy stabilizes and the proletariat may once again find work (Marx, 1844: 241, 258-259). Nevertheless, the greater deprivations endured by workers during crises become increasingly apparent to them and underline the basic differences between themselves and the owning class.

Of course, some capitalists also suffer greatly in hard economic times. This brings us to the fourth process in capitalist development that helps to bring about its transformation. It is mainly the small-scale capitalists— what Marx calls the *petty bourgeoisie*—and not the large-scale owners who are most vulnerable in economic crises. Small owners have fewer reserves of wealth for absorbing losses and less influence for getting bank credit or government subsidies than do large owners. Thus, in economic crises, small-scale capitalists increasingly are forced to sell their businesses or else go bankrupt and are taken over by the larger capitalists. As Marx puts it, the large-scale owners gradually "squeeze out the smaller capitalist" (Marx, 1844: 252; also Marx, 1844: 238). It is partly in this way that the economy stabilizes, because the overproduction problem lessens as the number of capitalist producers shrinks. In addition, however, the decline of the petty bourgeoisie means a concentration and centralization of ownership in the hands of an ever-smaller pool of large-scale capitalists. The dispossessed petty bourgeoisie and their descendants "sink gradually into the proletariat" (Marx and Engels, 1848: 71) and the ranks of the propertyless working class increase substantially (see also Marx, 1844: 258; 1867: 625–628). At the same time, the eventual disappearance of the intermediate class of small owners makes it more and more apparent that workers and big capital are the only two classes of consequence in the social structure. This realization promotes the awareness among workers of the clear disparity between the two classes, of the separation between bourgeoisie and proletariat. The process whereby this separation emerges and widens is referred to as *class polarization*.

The fifth feature of capitalism to consider here is its alleged tendency to homogenize the working class as time goes on. In the early stages of capitalism, the proletariat is a varied and divided collection of people with different skill levels. These divisions—between skilled artisans and unskilled manual labourers, for example—are obstacles in the way of proletarian solidarity, since the common interests of workers are less apparent when they perform different tasks and receive different wages. Because of the continual technological advances and mechanization of production as capitalism matures, these distinctions become blurred. Increasingly, all proletarians become the same type of worker: semiskilled minders of machines. This common position is an added impetus toward proletarian class awareness (Marx, 1844: 237; Marx and Engels, 1848: 71).

A sixth process—one that stimulates class polarization still further—concerns the gap in material well-being and living standards between capitalists and workers. As capitalism matures, *absolute* living standards for workers may improve, or at least not worsen, because of the huge productive capacity of capitalism and the surplus wealth this production generates. However, the *relative* difference in economic rewards between owners and workers continues to widen, in Marx's view, because a larger share of the surplus always goes to the capitalists (Marx, 1867: 645). Because of this widening gap and the relative impoverishment of the workers, class polarization once again is promoted.

A seventh and final factor in capitalist development that spurs its overthrow is the rise of stock ownership and the joint-stock company. As noted earlier, the capitalist class begins by playing a progressive role in social change. Its innovations and leadership are essential in the push toward the higher stages of human society that Marx envisions, and its members play an active part in managing the productive process and making administrative decisions. However, under advanced capitalism, businesses increasingly are owned only on paper, with individuals buying stocks in companies. To Marx, such a system makes it patently obvious to the workers that mere ownership of property is entirely superfluous to productive activity. Owners, as stockholders, control the means of production and accumulate rewards from it, yet offer little in return. They may not even be involved in administration, since they can hire managers to perform this task for them. Once again, then, the distinction between proletarian and capitalist is magnified and the dispensability of the bourgeoisie underscored (Marx, 1894: 427–429).

Obstacles to Revolution: Ideology and Superstructure

All of these processes and developments in the capitalist system promote the emergence of the fundamental division between bourgeoisie and proletariat. The objective differences in living conditions and the basic opposition of interests are there. Nevertheless, they must be recognized as such by the proletariat before revolutionary change can occur. The proletariat must then organize, mobilize, and act as a class-conscious group to transform society.

One might expect Marx to consider both the recognition of the situation by workers and subsequent revolutionary action to be unavoidable occurrences. However, except for some of his more polemical statements on the inevitability of revolution (e.g., Marx and Engels, 1848: 79), his actual position is more cautious, though optimistic. Marx's views here reflect his awareness that there are several important obstacles blocking the path to the overthrow of capitalism. These obstacles are sometimes classified under the general term, *superstructure*. Superstructure is frequently a confusing concept because it can refer to two separate but related phenomena. First, the superstructure is bound up with the concept of *ideology*, which is essentially the set of ideas or beliefs that govern and guide people's lives. But superstructure can also designate the actual social structures or organizations that people erect to implement these ideas. Thus, for example, religion is a system of beliefs about the world, while the church organizations themselves are the structural embodiments of these religious ideas. The state or political structure is likewise the structural manifestation of the laws, political enactments, rules, and regulations by which society is governed. In many ways, the superstructure concept is like the familiar sociological concept of institution, which can take on a similar dual meaning (Williams, 1960: 30–35, 515–520).

Marx seems to use the concept of superstructure in both these senses and rarely distinguishes between them. The state and religion are the major components of the superstructure in Marx's conception, but all "ideological forms," including "the legal, political, religious, aesthetic, or philosophic," are involved (Marx, 1859: 52; 1867: 82). Thus, any structure that serves as a concrete representation of the ideas that guide or govern us, and moreover is used as a creator and sender of these ideas, is presumably part of this ideological apparatus or superstructure. Hence, the educational system and media of mass communication would be included here, for they play an increasingly important role in modern society as means by which people learn ideas and acquire information (see Marx and Engels, 1848: 88).

The term, superstructure, implies that some substructure, or *infrastructure*, also exists. To Marx, this underlying basis to social life is of course the economic system, the "mode of production of material life" (Marx, 1859: 52; 1867: 82). One way to conceive of these two ideas is to think of the superstructure as somewhat like the visible parts of a plant, which exist above ground but grow as they do largely because of the operation of the roots lying beneath the surface, i.e., the economic infrastructure that nourishes and sustains the entire, interconnected, system. Thus, what Marx means to convey in the distinction between superstructure and infrastructure is not that the two types of phenomena are unrelated. On the contrary, he is strongly critical of Hegelian philosophy for setting up such an artificial separation. Even worse, in Marx's view, is that Hegel then assigns causal priority to ideas over material life, implying that ideas somehow exist before the human reality in which they take shape and are implemented. Marx takes such pains to reject this view that he sometimes seems to make the opposite, and equally tenuous, argument that ideas are totally secondary results of social or economic conditions and have no part in shaping these conditions (e.g., Marx, 1867: 19). In fact, Marx's position is for the most part a middle one, somewhere between the *materialist* and the *idealist* view (Avineri, 1968: 69). Ideas and the material conditions in which they operate clearly interact with one another in a continual process of change, and the existence of one is not possible without the other. To argue that ideas in themselves can have no causal impact on material reality is to contradict the great influence of Marx's own ideas on the course of historical social change (Berlin, 1963: 284).

The distinction between ideas and material reality is relevant to our discussion, because both the ideological and the structural aspects of the superstructure are useful to the capitalist class for maintaining the existing economic system, or infrastructure. To Marx, it is clear that the dominant ideas in any historical epoch are the ideas that the dominant class in that epoch generates (Marx and Engels, 1846: 59). Dissenting ideas may find the occasional outlet, as in Marx's own radical writings as a journalist and author. However, the values, beliefs, and styles of thought that predominate are mainly those created by the ruling class. In capitalism, then, it is the bourgeoisie's values and ideas that are most widely heard and accepted by the population—not only by the members of the bourgeoisie but also by large portions of the working class. This is one of the subtler and more insidious ways in which the path to revolution is obstructed. The bourgeois overthrow of feudalism was for some time impeded by the fact that the bourgeoisie itself accepted or believed in loyalty to higher authority, the divine right of kings, and other tenets of feudal society. In the same way, workers learn through the ideological apparatus of capitalist society that freedom, individualism, and equality are the guiding beliefs of the modern age. The power of ideas is evidenced by the belief among so many workers that these ideas really do operate in capitalist society, even though to Marx they only serve the interests of the privileged class. To Marx, the workers are free only to sell their labour power to the bourgeoisie and hence are equal only in their universal exploitation by the capitalist class (Marx and Engels, 1846: 39–41; 1848: 84–85; Berlin, 1963: 149).

All the segments of the superstructure—the political, legal, religious, educational, and communication systems—disseminate these views that actually favour the dominant class. In addition, certain branches of the political or state apparatus also serve as the "organized force of society" (Marx, 1867: 751). That is, the military and the police play a special coercive role as enforcers of the ideas, especially the laws, enacted and administered in capitalist society. Marx asserts in the *Communist Manifesto* that the state is a mere "committee for managing the common affairs of the whole bourgeoisie" (Marx and Engels, 1848: 61). This is probably a polemical exaggeration by Marx, in which he purposely understates the power of the state under capitalism. Nevertheless, Marx does believe that the prime purposes of the political system in capitalism are to administer and implement the legal relations of the society and to use force if necessary against those who disobey these laws. At the same time, to the extent that the laws are consistent with bourgeois interests for the most part, then the actions of the state are ultimately in the interests of the bourgeoisie. In this way, the state leadership can appear to be acting for the general good and enforcing the law in the public interest. State leaders may even believe sincerely that they are doing so, since politicians are equally subject to the seductive appeal of the dominant ideology. Yet, in seeming to serve the general good, the state ultimately works in the overall interest of those who control the economic structure—the bourgeoisie (see Jessop, 1982: 7–20; Sayer, 1991: 72–76).

Capitalism's End and Future Society

With all these resources at the disposal of the capitalist class, one may wonder why Marx expected a proletarian revolution to occur at all. Indeed, some critics would point to the absence of any current worldwide socialist movement as evidence that Marx's expectations will go unfulfilled. At the same time, however, many of the problems and developments Marx foresaw in the capitalist system have occurred, especially the regular economic crises. This suggests that his views have been at least partly confirmed.

As we have noted, Marx sees the end of capitalism coming about because of the internal problems it generates for itself. He expects the class polarization resulting from these difficulties to override gradually the ideological obstacles standing in the way of the working class's complete awareness of their common position and their revolutionary potential. Marx argues that the proletarians in capitalism share a common class position by the very fact that they are all people excluded from control of the means of production. In Marx's terms, the proletariat forms in this sense a "class in itself" (Marx, 1847: 211). The awareness of this common objective position, coupled with a belief in the possibility of change and a concerted desire to seek change, will, despite the repressive power of the state's coercive forces, convert this "class in itself" to a "class for itself"—a revolutionary force that will fundamentally alter capitalist society (Marx, 1847: 211).

Marx, however, is not confident in the proletariat's ability to accomplish revolutionary action without the added assistance, organization, and leadership of others (Avineri, 1968: 63). He mentions in particular "a small section of the ruling class," presumably enlightened members of the bourgeoisie and revolutionary intellectuals like himself, who recognize the destiny of the proletariat and go over to its side (Marx and Engels, 1848: 74–75).

Socialism and Dictatorship of the Proletariat

The specific details of how this revolutionary act will come about are never offered by Marx, for, as has been frequently pointed out, Marx is not concerned with providing a blueprint for the revolution or for the structure of the society that will supersede capitalism (Selsam, Goldway, and Martel, 1970: 20; Singer, 1980: 59). Nevertheless, some idea of what Marx envisions for the future can be gleaned from various excerpts of his writings.

It appears that Marx expects at least two stages in the system that will replace capitalism. The first stage, which may be termed *socialism,* is a temporary one that does not represent the best or highest form of society but is necessary in the transition from capitalism. During this transition, certain residual features of bourgeois society must be retained for a time, particularly some form of state or political apparatus. The state is needed to implement key changes that will eventually make its own existence unnecessary. Political power and decision making will be centralized in this body; however, unlike the state in capitalism, the socialist state will truly represent the interests of all people and not the special interests of any single group.

This "dictatorship of the proletariat," as Marx sometimes refers to it (Marx, 1875: 16), will ensure that important initial policies are instituted: abolition of landed property holdings and inheritance rights, centralization of banking in a government bank, institution of a graduated income tax, partial extension of the state ownership of factories and production, abolition of some forms of child labour, and free education for children (Marx and Engels, 1848: 94). Many of these ideas seem quite moderate by contemporary standards and, in fact, have been put into practice by most present-day capitalist democratic governments. The suggestion of only a partial movement of the state to take over production in this first stage seems especially restrained (Avineri, 1968: 206), as does the idea that not all forms of child labour will be eliminated.

However, it is clear to Marx that ultimately the entire productive process must be taken out of the hands of private interests. Marx sees the state-directed socialist system as an initial stage that is required while transitional wrinkles are ironed out. During this period, the truly social or communal nature of humanity and the system of material production will gradually become clear to all. The extraction of surplus by a dominant group for its own purposes will no longer occur. Instead, wealth generated in production will be distributed to all workers, after a common fund has first been set aside to provide education, health facilities, social services, and the like. The distribution of wealth during this first stage of socialism will be done on the basis that each worker "receives back from society—after the deductions have been made— exactly what he gives to it" (Marx, 1875: 8; see Giddens, 1971: 61). This policy suggests that all people will be rewarded, but not necessarily equally, since individuals will clearly vary in terms of the ability and effort they put into their work. This system, in Marx's words, "tacitly recognizes unequal individual endowment and thus productive capacity as natural privileges" (Marx, 1875: 9).

This clearly is not a classless society, at least in the sense that all people are treated exactly the same. Again, however, Marx sees this as a transitional phase in the move toward the final system. In this first stage, people have yet to overcome all the divisive tendencies instilled in them by the ills of capitalist ideology, all the defects of a changing system that "has just emerged after prolonged birth pangs from capitalist society" (Marx, 1875: 10). Marx believed that "it was in general incorrect to make a fuss about so-called '*distribution*' of goods" (Marx, 1875: 10). The most important issue for Marx is *relations* between groups, not the *distribution* of material goods or wealth to individuals. Marx sees the concern with such issues, including disputes over unequal incomes among people, as a problem stemming from the old bourgeois consciousness and its emphasis on individualism. Once the truly social basis of production is recognized, once it is clear that work must be a cooperative enterprise involving every worker and in the interest of every worker, then concerns with the dis- tribution of wealth will dissipate. Differences in ability, effort, responsi- bility, and so on may continue but will be greatly reduced. People will lose their view of one another as competitors and instead find satisfaction in their differences and in the different but mutually supportive roles they play in the collective process of production (Marx, 1847: 190; see Avineri, 1968: 232).

Here Marx seems to allow for the continued existence of a division of labour into specific tasks for each worker, despite the fact that elsewhere he frequently points to the evils wrought in capitalism by the division of labour (e.g., Marx, 1847: 182). It would appear that what will happen as society progresses through the first stage of socialism is a gradual revision and transformation of the division of labour. Two key changes in particular may be noted. First, there will be much greater diversity in the jobs that each worker is able to do. Work may be subdivided into specialized segments, but the same worker will not always perform the same segment. Each worker will be able to "hunt in the morning, fish in the afternoon, rear cattle in the evening, criticise after dinner" (Marx and Engels, 1846: 47). This seems to be Marx's quaint way of saying that in the future work will be more varied and workers more versatile. Second, it appears that Marx foresees the expansion of mechanized and automated production methods as a means to eliminate the routine drudgery imposed by the division of labour under capitalism. The "progress of technology" and "the application of science to production" under socialism will mean that the worker will be a "regulator" of the production process, "mastering it" rather than being enslaved by it (Marx, 1858: 705; also Marx, 1847: 190; see also McLellan, 1971: 216; Giddens, 1971: 63).

Communism

When these changes in the consciousness of people and in the structure of society are engendered and allowed to flourish, the first stage of socialism will have begun to fade, bringing on the highest societal form, true *communism*. The state's role as central decision maker will have been played out, and the state "will have died away" (Marx, 1875: 16). It will be replaced by a decentralized, nonauthoritarian system of administration of the society by all the people. It is not clear how this new form of administration is to be organized and operated—once again there is no detailed plan offered by Marx—but the new human consciousness, which sees people in their true character, will ensure that the society will function unencumbered by the distorting effects of bourgeois self-interest, greed, and the pursuit of power. Then it will be possible to implement Marx's famous dictum of equality and selflessness: "From each according to his ability, to each according to his needs" (Marx, 1875: 10).

It is essentially Marx's faith in the potential of humans to act as social beings in a universal and truly equal society that is the basis for his expectations and visions of the future of society. The view of human nature and the belief in the perfectibility of people that Marx acquired early in life, and that he retained throughout his intellectual development, are reflected here, although they seem strangely paradoxical, given his own apparent aloofness toward humanity in general and his ultimate reluctance to involve himself directly in the revolution (Berlin, 1963: 1–3; Avineri, 1968: 251–252).

We will see in the subsequent chapters of this book the tremendous influence that Marx's ideas still have on the study of social inequality and class structure in contemporary societies. We will also see the skepticism that Marx has provoked in some of those who were to write after him. Both adherents and critics alike, however, readily acknowledge the deep intellectual debt owed to Marx. His analysis of class structure in capitalist society provides a touchstone against which to assess and compare all subsequent views of social inequality in advanced societies. In the next chapter, we shall consider the next great classical theorist of social inequality, Max Weber. As we shall see, Weber provides a prime example of how both the positive influences of Marx's theory and numerous critical reactions to specific aspects of it can be bound up in the same writer's work.

SUMMARY

In this chapter, we have reviewed the key elements in Marx's theory of class and his analysis of capitalism. We began with an outline of some of the significant experiences in his own life that helped shape his thought. We then considered his contention that class struggle has been the consistent basis for social change and inequality throughout history. The remainder of the chapter dealt with the key issues Marx raises concerning the capitalist form of society. We discussed the distorted nature of work under capitalism; the seven processes inherent in capitalism that Marx believed would promote its transformation from within; the various obstacles in the way of capitalism's revolutionary overthrow; and Marx's views about the ultimate end of capitalism, with a brief sketch of the socialist and communist societal forms that he anticipated.

CHAPTER THREE

Max Weber and the Multiple Bases for Inequality

> *In the last analysis, the processes of economic development are struggles for power.*
> MAX WEBER
> Inaugural Lecture, Freiburg University, 1894

INTRODUCTION

Max Weber is the second major classical theorist that we shall consider. Weber's analysis is frequently viewed as a distinct alternative to and departure from Marx's theory of class. Indeed, there are significant differences between the ideas of Marx and Weber that should not be overlooked. At the same time, however, we shall find notable similarities in the interests and conclusions of these two writers. A thorough appreciation of both the similarities and the differences in their work is essential to our understanding of social inequality and of the current state of theory in this field of study.

Perhaps the most obvious similarity between Marx and Weber is their common concern with examining the origins and development of modern capitalist society using an historical method. In this regard, of course, Marx and Weber share with many nineteenth-century social analysts an interest in how societies have changed or evolved from traditional to modern forms. The two writers are not in total accord about the nature and extent of social inequality, the forces that govern social change, or the likelihood of inequality in future societies. Nevertheless, there can be no doubt that Weber's general conceptions of capitalism, social class, and numerous related ideas are greatly influenced by Marx's writings, virtually all of which preceded his own work. Weber himself identifies Marx as one of the two major intellectual influences (along with Friedrich Nietzsche) of Weber's time (Gerth and Mills, 1967: 61–62; Giddens, 1972: 58; Coser, 1977: 249–50). In fact, as Erik Olin Wright (1997: 29–30) has argued, there are some aspects of Weber's analysis of inequality, especially his discussion of the key concept of class, that suggest he often speaks "in a Marxian voice."

Sometimes Weber's differences with Marx are less about substantive issues in sociology and more about such matters as the divergent political positions of the two men and their distinct opinions on the likelihood or desirability of socialist revolution. Moreover, often Weber's alleged differences with Marx are really disagreements with certain Marxists of Weber's day, disciples of Marx who interpreted his theories and applied them for their own purposes (Löwith, 1982: 101). Even in these cases it is fair to say that a good deal of Weber's work is a "positive critique" of Marx and Marxism, a "fruitful battle with historical materialism" (Gerth and Mills, 1967: 63; see also Sayer, 1991: 3–4).

One of the unfortunate parallels between the two writers is that neither one provides a systematic and detailed analysis of social inequality in modern societies. This omission is partly because neither man lived long enough to complete this task and partly because neither thinker was concerned with the issue of social inequality as his principal focus. It is somewhat ironic that theoretical developments in the study of social inequality owe so much to two writers for whom the topic was not really their primary interest. For Weber as for Marx, then, we must rely on rather short discussions, scattered through his writings, as the means for deciphering his general view of social inequality.

In our analysis of Weber's thought we will take the position that his major contribution to the theory of social inequality lies in his attempt to offer a positive or constructive critique of Marx's ideas, especially as these ideas were interpreted by Marxists of Weber's time. In general, Weber's treatment of Marxist thought can be characterized as "critically respectful" (Collins, 1986a: 37). Because of Weber's advantage in living after both Marx and the early Marxists, he is able to draw on a wider and more comprehensive range of sources and evidence, all of which lead him to propose a number of modifications to the Marxist view (Collins, 1986a: 38–40).

If there is a common theme in Weber's criticisms, it is that the accurate description and explanation of inequality and other social phenomena involve much greater complexity and variability than some Marxists seem to suggest. This is not to imply that Marxism in its many forms is crude and simplistic, although this charge can be made against specific versions of social theory, Marxist or otherwise. Rather, it is to say that Marxist theory proceeds from the conviction that there is a single, ultimate basis upon which social life is built: the economic or material realm of human activity. Other social phenomena—the political and religious, for example—may be worthy of study, but, in the last analysis, these too are explicable in terms of economic forces. To Marxists, then, the study of these aspects of society should not be allowed to cloud our understanding that the real basis of social structure and inequality is economic.

For Weber, however, the range of noneconomic social forces cannot be dismissed as merely secondary to, or completely determined by, the economic. On the contrary, noneconomic considerations, especially the ideas and interests that emerge from politics, religion, and other institutional structures, have a certain autonomy from the economic in many instances. Moreover, these forces sometimes influence economic structures and behaviour as much as they are influenced by them. To many Marxists, such an emphasis on multiple structures only distorts our understanding, because it lays down a smoke screen that masks the underlying material basis for social life. In the Weberian view, these details do not distort our picture of reality. On the contrary, distortion stems from failing to recognize these diverse explanations and multiple causes for social inequality.

A detailed treatment of this general difference between Weber and Marxism, as well as an assessment of the common elements that the works of Marx and Weber share, is the central concern of this chapter. We begin with a brief examination of the life experiences and intellectual influences that affected Weber's thought.

BIOGRAPHICAL AND INTELLECTUAL SKETCH

Early Years

Max Weber was born April 21, 1864, at Erfurt, in the German province of Thuringia. By this time, Karl Marx was a 46-year-old émigré living in England; hence, the two men never met. There are some broad similarities between the backgrounds of Marx and Weber. Like Marx, Weber grew up in a relatively privileged, middle-class German household. Weber's father, like Marx's, was trained in law, although he eventually pursued a career in politics and government at the municipal and later the national level. Like Marx, Weber initially followed his father's wishes and enrolled in law school. This was at the University of Heidelberg in 1882, one year before Marx's death. In the same way that Marx spent his early university days carousing and having a good time, so too did Weber enjoy much of his initial student life. He joined his father's old fraternity and took a regular part in fencing bouts and drinking events. Thus, Weber, who had been a sickly, frail, and bookish child (Weber, 1926: 32–33, 45), developed into a hearty, somewhat barrel-shaped young man, complete with duelling scars (Weber, 1926: 70–71; Gerth and Mills, 1967: 8; Coser, 1977: 236; Käsler, 1988: 3). At the same time, Weber managed to be an excellent and conscientious student who was popular with his fellows, partly because of his willingness to help them at examination time.

In contrast to Marx, Weber chose to stick with his legal studies, although, like Marx, he became widely read in other areas, including economics, philosophy, history, and theology. His first year in Heidelberg was followed by a year of compulsory military service at Strasbourg, after which he resumed his legal studies, this time at Marx's old school, the University of Berlin. The move to Berlin was prompted partly by the wishes of Weber's parents, who had settled there some years before. They were concerned about his boisterous behaviour at Heidelberg and sought to introduce more discipline into his life by having him live at home (Käsler, 1988: 5).

Family Influences

The family emphasis on discipline was particularly evident in Weber's father, a strict, authoritarian, and often brutish man who at times mis-treated his wife. One possible effect on Weber of this family atmosphere was a rejection of arbitrary power and a distrust of authority without accountability. The divergence between Weber and his father is evident in his dislike for his father's political beliefs, which favoured the con-servative, reactionary policies of the German Kaiser and of Chancellor Otto von Bismarck. Like Marx, Weber was a supporter of democracy and human freedom. This is important to recognize because, as Giddens notes, some writers have mistaken Weber's patriotism for Germany and his belief in strong leadership as evidence of right-wing or fascist tendencies (Giddens, 1972: 7–8). However, in contrast to Marx, Weber was far less optimistic about the prospects for democracy's survival in future societies. Moreover, despite sharing Marx's interest in the plight of the working class (Bendix and Roth, 1971: 15–17, 233; Löwith, 1982: 22) and despite revealing occasional socialist leanings at times (Weber, 1926: 630), Weber ulti-mately rejected socialism as the means of achieving these goals. On the contrary, he suspected that socialism would be an even greater threat to democracy and freedom than would capitalism. Humanity's best hope, according to Weber, lay in a liberal political system, guided by strong but enlightened leaders and operating within an essentially capitalist eco-nomic framework. Weber's political views were influenced in this direction by his uncle, Herman Baumgarten, whom he visited frequently during and after his military service at Strasbourg and who retained a strong belief in the liberal democratic principles that Weber's father had forsaken.

Weber's visits with his uncle and other relatives in Strasbourg led to other events that significantly influenced his thought. Although Weber's mother was a devout Protestant, her religious beliefs had not made a notable impression on Weber in his early years. However, in Strasbourg, Weber was in contact with several family members, including his mother's sisters, some of whom were prone to religious and even mystical experiences (Weber, 1926: 82–87; Gerth and Mills, 1967: 9). Thus, Weber witnessed first-hand the power that religious ideas and beliefs can have over people. Though Weber never became a religious man himself, these events helped stimulate his eventual interest in the study of religion and the influence that religious beliefs and other ideas could have in shaping society. Weber's position on this point may be contrasted with that of certain Marxists of his day, who argued that ideas and beliefs are wholly products of social interaction and organization, especially within the sphere of economic production. Weber disagreed with those who adopted this materialist view in the extreme, thereby denying the possibility that ideas could themselves influence and even generate economic structures and behaviour, rather than being mere consequences of these material forces.

Success and Crisis

Weber lived at his parents' suburban Berlin home for almost eight years, beginning in 1884. During this period, he revealed both tremendous scholarly ability and a voracious appetite for work (Käsler, 1988: 5, 11). He completed law school, worked as a junior barrister, and at the same time followed a rigorous course of study leading to a Ph.D. in law in 1889. Unlike Marx, who was denied an academic career because of his political beliefs, Weber enjoyed a meteoric rise in the university community. He began as a lecturer in law at Berlin in 1891. This position was soon followed by a senior professorship in economics at Freiburg in 1894. In 1896, at age 32, Weber became Chair of economics at Heidelberg.

These successes, however, came at the price of a heavy workload that taxed him mentally and physically. Then, in 1897, a crisis triggered some major changes in Weber's life. During a visit by his parents to his home in Heidelberg, Weber argued bitterly with his father over the continued mistreatment of his mother and drove his father from the house. Not long after, Weber's father died. The guilt Weber felt over these events, coupled with the strain of overwork, led to his psychological collapse (Käsler, 1988: 10–12). Other accounts suggest that Weber's problems also stemmed in part from his dislike for the puritanical tendencies of his mother (Collins, 1986b: 20; see also Portis, 1986). In all events, Weber's mental condition was never completely diagnosed and recurred intermittently for the rest of his life. Until just before his death, Weber did not hold another full-time teaching position. Nevertheless, he managed a partial recovery and, with regular pauses for travel and recuperation, was able to continue his research. In fact, his most famous and significant works began to take shape only after this initial breakdown.

America and Capitalism

An important surge in Weber's work occurred soon after he travelled to the United States in late 1904. His first-hand observation of American capitalist society in action made a lasting impression on his thought in a variety of ways. While his reactions were not entirely positive, on the whole Weber admired the United States. He saw in its mass political parties, voluntary citizens' organizations, and other institutions the possibility that freedom and democracy might be sustained in future societies. Weber was a nationalist concerned with the development of his own country, and some have implied that he perceived in the United States the model for a new German society (Gerth and Mills, 1967: 17). However, it appears that Weber saw too many contrasts between America and Germany for any direct imitation to be practicable. Moreover, certain aspects of the American system underscored for Weber the paradoxical and even contradictory nature of mass democracy. In particular, Weber noted that bureaucracy—in the form of extensive political-party machines managed by professional politicians and organizers—was at the same time both essential to democratic action in large-scale complex societies and a threat to the democratic principles of equality and participation for all.

Weber was also impressed by the large business corporations in the United States. These enterprises were then coming to the forefront as the dominant forces in the world economy. The movement toward bureaucracy was apparent in the corporations, as well, contributing to Weber's ultimate conclusion that bureaucratization is a key process in the general trend of modern societies toward "rationalization." Weber perceived in all spheres of the social structure—politics, economics, religion, education, and so on—this tendency to develop permanent, organized systems for processing problems and people in regular, routine ways. Weber saw bureaucracy as the only organizational form capable of keeping modern, complex societies in operation, making it both inevitable and inescapable. Yet, at the same time, he regretted bureaucracy's destructive impact on the quality of human interaction and on human freedom (Bendix, 1962: 7, 458–459).

After his American travels, in the period up to World War I, Weber made a notable intellectual recovery. His writing began again, and in 1905 he published perhaps his most famous work, *The Protestant Ethic and the Spirit of Capitalism*. We have already discussed the Protestant background of Weber's mother, the religious experiences of his mother's family, and the subsequent impact of religion on Weber's thought. This impact is most evident in *The Protestant Ethic*. In this analysis, Weber argues that the historical development of modern capitalism was significantly influenced by the ideas and beliefs of certain ascetic Protestant sects, particularly Calvinism. This thesis is consistent with his general critique of materialism, for he seeks to demonstrate how ideas, especially religious beliefs, can and

Max Weber, 1864–1920

©1896–97 (b/w photo), German Photographer, (19th century)/Private Collection, Archives Charmet/Bridgeman Art Library.

do influence economic and social structures like capitalism, and are not simply reflections or products of such structures. His views on this point were also affected by his visit to America, where he observed the tendency for those of ascetic Protestant backgrounds to be overrepresented among the most successful capitalists (Weber, 1926: 326; Coser, 1977: 239).

Academic and Political Prominence

Weber's academic reputation now began to grow anew. His home in Heidelberg became a regular meeting place for prominent German intellectuals. He was soon writing numerous works on a variety of topics, including several essays on method in the social sciences. His key work for the study of social inequality, *Economy and Society,* was begun around this time (1909) and was continued intermittently but unfortunately was never finished (Aron, 1970: 305).

One crucial reason for the interruption of Weber's research was the outbreak of World War I in 1914. A patriotic German, Weber initially supported the war effort, enlisted as a reserve officer, and served as director of military hospitals around Heidelberg. However, he soon became disillusioned with the war, characteristically, because he questioned the motives and competence of the political regime of the German monarch, Kaiser Wilhelm II (Giddens, 1972: 21–22).

Weber had always been interested in his country's political fortunes, and he gradually became directly involved in them. He unofficially sought to convince the German leadership to stop the fighting, without success, and later acted as an adviser to the German delegation to the 1919 Versailles peace conference, which ended the war. His political activities also included participation in the drafting of a new German constitution and an increasingly large role in political campaigns. Some even viewed Weber as a potential candidate for the German presidency (Coser, 1977: 241). However, his earlier criticisms of the monarchy and of the Kaiser's conservative government made support from this quarter unlikely. His dislike of the opposition socialist factions—some of whom he suggested should be in either "the madhouse" or "the zoological gardens"—was also well known (Giddens, 1972: 25, 17). Thus, Weber's political prospects eventually dissipated.

Instead of taking up a political career, Weber remained with his academic pursuits and was finally able to resume full-time teaching, this time at the University of Munich (Bendix, 1962: 3). It was during this period, as well, that Weber engaged in a series of important and influential lectures, including debates with contemporary Marxists on the nature of capitalism. One series carried the characteristic title, "A Positive Critique of Historical Materialism" (Alexander, 1983: 55; see also Collins, 1986a: 19, 38).

His scholarly eminence and intellectual vigour reestablished, Weber seemed on the verge of even greater things when, in June 1920, he was stricken with pneumonia and suddenly died, at the age of 56. As with Marx, then, death cut short Weber's career before he was able to complete a thorough and systematic analysis of social inequality. As with Marx again, though, it is possible to piece together the major strands of Weber's work on this topic and to identify a more or less coherent basis for his emerging theory. A presentation of these main features of the Weberian perspective and how it compares with that of Marx is our next consideration.

THE WEBERIAN PERSPECTIVE: COMPLEXITY AND PLURALISM IN INEQUALITY

In the preceding chapter, we noted that the central point in Marx's theory of social structure is his conviction that, in the last analysis, societies take shape and change through one key process: the struggle between groups for control of the economy, the system of material production. To recognize Marx's predominant emphasis on economics is to deny neither his theoretical sophistication nor his obvious awareness of the complex nature of modern societies. Nevertheless, a tendency does exist to view Marx's position as an oversimplified one. In part, this tendency may be because subsequent analysts have been distracted from Marx's complete argument by certain polemical, and hence purposely exaggerated, statements in some of his writings, most notably *The Communist Manifesto*. In other instances, the responsibility for oversimplifying the original Marxian formulation falls on the shoulders of certain Marxist disciples, including some who have attempted to put socialist principles into practice in real societies (Avineri, 1968: 252; see also Sayer, 1991: 4).

In either case, it is clear that Weber's alternative view of social structure stems in part from his reaction to an oversimplified or "vulgar" Marxism. Thus, a consistent feature of Weber's sociology is his conviction that social processes and forces are always complex and rarely explicable in simple terms (Bendix, 1962: 6). Weber's emphasis on complexity is evident in a variety of ways that, taken together, form the substance of most of our discussion in this chapter. There are six areas in particular that we shall examine because of their relevance to the understanding of Weber's general perspective and his views on social inequality.

First, there is Weber's overall approach to research and sociological method, with its stress on *causal pluralism* and the *probabilistic* nature of social explanation. Second, there is Weber's sense of the often intricate interplay between ideas and material reality, or between the subjective and objective aspects of social life. The third point concerns Weber's multiple conception of the class structure in capitalist society. Fourth is Weber's pluralist view of the bases for hierarchy and group formation in social structures: class, status, and party. The fifth way in which Weber adds complexity to the analysis of inequality is in his outline of the several types of *power* or *domination* in social life. Finally, we shall assess Weber's contention that the control of economic production is just one means by which some people are able to dominate others. Thus, to understand social inequality, we must recognize the several *means of administration*, especially in the form of rationalized bureaucracies, that wield power in modern societies. This last point leads into the concluding discussion, in which we consider Weber's assessment of the prospects for democracy and the socialist alternative in future societies.

Weber on Method: Probability and Causal Pluralism

Max Weber wrote extensively on the topic of research methods in the social sciences (see especially the collection of essays in Shils and Finch, 1949). While it is beyond the purposes of this discussion to consider Weber's methodological orientation in detail, there are certain aspects of his position that are relevant to our analysis. To begin with, we can briefly characterize Weber's approach as a compromise between two opposing schools of thought: those who reject any possibility of applying the techniques of natural science to predict or explain social behaviour and those so-called positivists who, on the contrary, contend that the methods and assumptions of sociological research should be essentially identical with those of, say, physics or chemistry (Runciman, 1978: 65). Weber locates sociology somewhere between the natural sciences and a discipline like history, in terms of both the generality of explanation and the precision of research techniques. Whereas history concerns itself with the understanding of "important individual events," sociology deals with the observation and explanation of "generalized uniformities" in recurring empirical processes (Weber, 1922: 29, 19). Unlike history, sociology can offer theories to account for these empirical regularities. However, social explanations typically will not hold without qualification for all times, all cases, or all conditions. This feature makes sociological explanations distinct from those in natural science to the degree that the latter will hold invariably, assuming appropriate controls are applied to account for extraneous influences or circumstances.

There are many possible reasons for this essential difference between natural science and social science, including the lesser precision of social concepts and the fact that explanations in social science are subject to constant revision over time, as conditions change (Runciman, 1978: 65). Thus, while chemical elements will interact today in the same way as they did centuries ago, individuals or groups in different historical and social contexts may not do so, and the social scientist must take this possibility into account. For these and other reasons, Weber raises certain restrictions concerning the application of the scientific method in social research. Two of these restrictions or complications are the probabilistic nature of social inquiry and the inherent pluralism in social causation.

Probability

Weber frequently refers to probability when discussing social phenomena (e.g., Weber in Parsons, 1947: 99–100, 118–119, 126, 146). In effect, Weber argues that X *may* lead to Y in some, or even most, instances but rarely without exception. This means that we can assess, and sometimes calculate, the probability that X is a cause of Y, but we cannot establish that X will *always* lead to Y, since this is not in the nature of social processes (see Weber, 1922: 11–12; Giddens, 1971: 149, 153). If we take, for example, Marx's contention that the dominant ideas in a society are the ideas of the ruling class, this may be true for some or most of the ideas that predominate in a society but probably not for every dominant idea we could examine. The exceptions do not mean, however, that Marx's claim about the origin of ideas is incorrect. Rather, such exceptions indicate that the claim is applicable part of the time and that, within a certain probability, we can predict correctly what ideas predominate in society by examining those generated by the ruling class.

Causal Pluralism

And what of those exceptional cases that do not fit our prediction or explanation? These must be examined to determine, where possible, the reasons for their deviation from our expectations. According to Weber, there are numerous other factors that inevitably are not taken into account when one poses a simple explanation for something. It is conceivable that these additional factors are responsible for many of the deviant cases; hence, it is essential to search for multiple causes for social phenomena, following a strategy of "causal pluralism" (Gerth and Mills, 1967: 34, 54, 61). This method seeks social explanation by means of "a pluralistic analysis of factors, which may be isolated and gauged in terms of their respective causal weights" (Gerth and Mills, 1967: 65).

If, for example, we wished to understand why some people are rich and others are poor, we might try to explain these differences in wealth in terms of the family class background of the people involved. Perhaps all rich people were born into wealthy families and all poor people were born into poor families, with no other factors playing a part in economic success or failure. In that case, we would be able to explain economic inequality perfectly on the basis of one factor: inheritance of wealth. However, in a complex society, it is probable that other factors also influence one's economic position, not just inheritance. Perhaps an individual's gender or race has an effect on economic well-being, independent of inherited wealth. Even physical beauty or athletic prowess might be related to economic position. Of course, not all of these possible influences would have the same degree of impact. Presumably, for example, inheritance, gender, or race would have a greater effect than beauty in most cases. But, whatever the result of the inquiry, it is imperative that the social scientist look for the several potential causes for social phenomena and assess their relative effects, large or small, by means of empirical observation.

Subjective Factors and the Idealism-Materialism Debate

Subjective Meanings and Explanation

Another one of the complexities that mark social life, and that to Weber must be considered in any social theory, is the subjective nature of human interaction. This point is related to our previous discussion of Weber's method, since these subjective considerations signify another important distinction between natural and social science. One of the advantages of natural science, and another reason for its greater precision, is that it need deal only with objectively observable and interpretable information. Physics and chemistry, for example, analyze the behaviour of inanimate objects or forces that, under controlled conditions, can be seen to behave in perfectly predictable ways. In social science, however, the units of observation are individual human beings, whose actions are determined not only by objective conditions but also by subjective forces that lie outside the realm of natural science. Thus, social explanations must take into account these additional subjective aspects of social life: the meanings people attach to their actions, the ideas that govern their behaviour, and their consciousness and perceptions of the world around them.

A related problem that is unique to social science is that human beings are thinking and reasoning social actors. This means that people's behaviour may often be difficult to predict with accuracy, because humans can consciously change their behaviour if they are made aware of the predictions made about them by social scientists (e.g., Giddens, 1982: 12–16). For example, a researcher's prediction that a serious disease is going to spread widely throughout the population could turn out to be incorrect, in part because that same prediction is made known to the population and causes many people to take precautions to avoid the disease.

Despite such problems, however, Weber suggests that certain kinds of human behaviour can be readily understood by analyzing subjective phenomena. What Weber calls *rational action*, the calculated pursuit of individual interests, is predictable with a high level of certainty because one can successfully assume what subjective motives are at work in most cases. This kind of behaviour is epitomized in the actions of capitalists in the marketplace, where subjective interests are geared almost completely to the profit motive or the maximization of wealth (Weber, 1922: 30).

However, not all human behaviour follows the ideal type of rational action. (*Ideal type* refers to a *pure* form here, not necessarily a desirable one.) Some actions are governed by *nonrational* or even irrational considerations—superstition, love, envy, vengeance, and so on—usually in combination with rational concerns. The more that human actions depart from the purely rational type, the more difficult it becomes to discern, in the external acts of people, the subjective intent of their behaviour (Weber, 1922: 6). This both adds to the complexity and decreases the precision of social research and social explanation. At the same time, however, this opens up for the sociologist a whole area of inquiry that is closed to the natural scientist and so, in this sense, is an advantage of sociology over natural science. In fact, to Weber, it is this special focus of sociology on subjective meaning and explanation that distinguishes it from other kinds of knowledge (Weber, 1922: 15).

Ideas and Material Life

Weber's stress on subjective factors as an integral part of social explanation is consistent once again with his rejection of vulgar Marxist theory. In this case, the point at issue primarily concerns the debate over the relationship between ideas and material reality. As noted in Chapter 2, one of Marx's early tasks was to demonstrate the fallacy in Hegel's idealist philosophy, in which ideas are analyzed without regard for the social conditions in which they emerge and operate. Despite the occasional remark by Marx to the contrary (e.g., Marx, 1867: 19), Weber and Marx appear in fact to hold similar positions on this point (Giddens, 1971: 209–210). A key problem for Weber, however, is that some of Marx's disciples seem to have replaced Marx's original formulation with an oversimplified and extreme materialist view that treats ideas and other subjective phenomena as mere byproducts of concrete economic conditions, and therefore irrelevant as explanations of human action.

Throughout much of his work, Weber takes pains to demonstrate that social life cannot be understood so simply. In his view, those who invoke only material, economic factors to explain social action are often likely to be wrong because the subjective meanings and ideas people live by frequently produce effects different from those that a simple materialist theory would predict. One of Weber's early works points out, for example, that peasants in nineteenth-century Germany chose the relative freedom (but economic hardship) of working as wage labourers over the greater economic well-being (but servitude) of bonded serfdom (Giddens, 1971: 122–123). To Weber, this is but one instance of the tendency for people at times to put a subjective meaning, the idea of freedom in this case, ahead of simple material conditions or interests when choosing a course of action.

Undoubtedly, the best-known work by Weber that addresses this issue is *The Protestant Ethic and the Spirit of Capitalism*. One of the objectives of this study is to show how "ideas become effective forces in history" (Weber, 1905: 90; see also Weber, 1926: 339–341). Thus, Weber examines the relationship between the religious ideas and beliefs of "ascetic" Protestant sects like Calvinism and the rise of modern capitalism. Weber observes the Calvinist belief that all people are predestined by God either to salvation or to hell. He believes that this makes it important to each Calvinist to be successful in this world, in whatever *calling* he or she chooses. The action of the Calvinist cannot of course determine whether he or she is saved or damned, since everyone is predestined. However, the psychological need to *appear* to be one of the chosen, to "shew oneself approved unto God" by being a success in this life, is sufficient motivation. This concerted drive for success, when combined with the belief in asceticism, which refers to the rejection of worldly pleasures, means that Calvinists tend to accumulate wealth that cannot be spent on frivolities and so is invested in ever-expanding capitalist enterprises. Thus, *part* of the reason why certain Protestant sects are overrepresented among the capitalist class, and *part* of the reason why capitalism took the specific direction of development that it did, can be traced to certain dominant ideas and beliefs promoted by these religions.

It is important to understand that Weber never argues that Calvinist religious ideas are the only reasons for capitalism's emergence or that capitalism would not have occurred without them. He explicitly rejects such a simplistic explanation as "foolish and doctrinaire" (Weber, 1905: 91; see also Löwith, 1982: 103). His later writings on the subject reveal an even clearer disavowal of strictly idealist explanations and, if anything, put greater stress on the importance of organizational or structural factors in the emergence of modern capitalism (Collins, 1980; Collins, 1986a: 19–37). It is notable, as well, that Weber is not alone in seeing at least some connection between ascetic Protestantism and the rise of capitalism. Sayer (1991: 100) has pointed out that Marx also voiced somewhat similar views at times, and linked the beliefs of such religions as "English Puritanism and Dutch Protestantism" with certain features of the capitalist system (see Marx, 1858: 164; 1867: 592–594).

Nevertheless, for both Weber and Marx, it would be a mistake to overstate this argument. Thus, as Weber contends, it contributes little to our understanding if we "substitute for a one-sided materialism an equally one-sided spiritualistic causal interpretation" (Weber, 1905: 183). One-sided explanations, whether they are based on economic or religious factors, are obviously inconsistent with Weber's sense of the complexity of social processes and with his deep-seated belief in causal pluralism. Instead, there is an intricate interplay between objective social conditions and the subjective meanings that people attach to their own actions. These forces *in combination* are mainly responsible for the patterns of human action and for the shaping of social structures (see also Weber, 1926: 332–335).

Multiple Classes in Capitalism

To this point, we have attempted to demonstrate the pluralist nature of Weber's assumptions about sociological theory and method. The same characteristics are evident in his views about the specific area of social inquiry of concern to us: social inequality. We can detect these characteristics first of all in Weber's conception of class, the central idea in Marx's analysis of social inequality.

In Chapter 2, we noted that Marx defines class by distinguishing between two key groupings: the owners of the means of economic production, or bourgeoisie, and those nonowners who must work as wage labourers, the proletariat. Weber's formulation is similar to Marx's definition in some ways, for both writers treat classes in the most basic sense as economic entities. One difference, however, is that Marx is concerned primarily with the social *relations* between his two classes in the productive sphere, especially the relations of domination and exploitation of the workers by the owners and the inherent conflict that these generate between classes. As we have seen, the simple *distribution* of wealth is a secondary consideration for Marx. Weber's emphasis is somewhat the reverse, at least with respect to the concept of economic class. Weber also speaks of class struggle and relations of class domination, but he places more emphasis than Marx on the distribution of valued objects and the process by which some people get more than others.

To understand Weber's view, we can begin by noting that he sees classes as economic categories, developing out of human interaction in a *market*. A market here is basically a system of competitive exchange whereby individuals buy and sell things of value in the pursuit of profit (Weber, 1922: 82). These things of value are in effect equivalent to what Weber calls "utilities," which include both material "goods," especially property, products, or possessions, and human "services," namely personal skills and labour power (Weber, 1922: 63, 68–69). Essentially, a class is simply an aggregate of people sharing common "situations" in this market and therefore having similar "economic interests" and "life chances" (Weber, 1922: 927–928).

We can see that Weber's initial discussion is similar to Marx's, because his definition leads to a simple distinction between those who have property and those who have only services to exchange in the market-place: " 'Property' and 'lack of property' are, therefore, the basic categories of all class situations" (Weber, 1922: 927). At this point, however, Weber follows his normal course, and reasserts his disagreement with most of his Marxist contemporaries, by pointing out what he believes are other important complexities that the simple two-class model hides: "Class situations are further differentiated . . . according to the kind of property . . . and the kind of services that can be offered in the market" (Weber, 1922: 928). Thus, the existence of different kinds of property makes for additional classes within the broad propertied group, and different types of services and job skills distinguish segments of the working class from one another. Taken to its logical extreme, this definition leads to difficulties because it implies that each of the various kinds of property people own, and each of the numerous job categories that can be identified in the marketplace, may signify a separate class (Weber, 1922: 928). Thus, almost every individual in a complex economic system like modern capitalism could in a sense represent a distinct class, and the concept of class would then have little use or meaning (Giddens, 1973: 78–79).

In actual practice, fortunately, Weber does not apply his formulation in this extreme manner. In the end, his version of the class structure lies between the simple dichotomy of the cruder versions of Marxism and the uncompromising pluralism that his own definition implies. Nevertheless, to present the Weberian view of the structure of classes in capitalism is still difficult, because of a disjuncture between two of the sections on class in his unfinished work, *Economy and Society* (Weber, 1922: 302–307,

926–940). These sections were written at different times and seem inconsistent in places, suggesting that Weber may have revised his initial view after some reconsideration (Giddens, 1973: 79).

One solution to these difficulties is to draw a distinction between the idea of class and the related but separate concept of *social class*. As we have already discussed, in Weber's terms a class is basically just a category, a set of individuals who happen to live in similar economic circumstances and have similar economic interests. Thus, "a class does not in itself constitute a group [or community]" (Weber, 1922: 930). Individuals in a simple economic class lack the sense of their common position and consciousness of their common interests that a true group or community possesses. It is when these subjective qualities are added to the aggregate in question and a process of real group formation emerges that Weber's idea of social class becomes operative. In effect, then, social classes are best understood as economic classes that have acquired in varying degrees some subjective sense of "unity" and "class-conscious organization" (Weber, 1922: 302, 305). Here again, the importance Weber attaches to subjective processes is quite evident.

At the risk of some oversimplification, it is worth pointing out that Weber's distinction between class and social class corresponds broadly to Marx's distinction between a class in itself and a class for itself. That is, just as Marx's class in itself signifies a collectivity that is identifiable on purely economic grounds, without reference to group awareness, so too does Weber's concept of class. And, just as Marx's class for itself possesses a consciousness of common position and interests, so too does Weber's social class. The major difference to be noted here is that Marx uses his formulation primarily with reference to the proletariat, in order to indicate the two stages through which the working class must move if it is to become a revolutionary political force for transforming capitalist society. In contrast, Weber's discussion is intended to delineate a range of social classes, with degrees of "unity" or group awareness that are likely to be quite different and "highly variable" (Weber, 1922: 302). Moreover, as we shall see, Weber believes that the highest degree of group consciousness and the most potential for political action and control lie not with the working class but with those at the top of the class hierarchy or social structure.

With this additional element of subjective awareness included, Weber's potentially countless economic class categories in capitalism tend to coalesce into a limited set of four social classes (but see Scott, 1979: 182). In effect, Weber's (and Marx's) distinction between the propertied and the propertyless is supplemented by the addition of two other classes (Weber, 1922: 305). What this addition amounts to, first, is separating the bourgeoisie into those who control large amounts of property, the big capitalists, and those who have relatively small amounts of productive property, the petty bourgeoisie. Similarly, the propertyless category is subdivided by Weber, this time according to the level of skill and training required of those who sell their services in the marketplace. The key distinction here is between the "working class," who tend to have manual labour power alone at their disposal, and those who have more marketable training and skills as "specialists," "technicians," "white-collar employees," or "civil servants" (Weber, 1922: 305). Thus, interposed between the large-scale bourgeoisie and the mass of proletarians, we have two middle classes: owners of small, independent shops, businesses, and farms; and a salaried nonmanual class with special education or skills in such areas as law, medicine, the sciences, accounting, office management or administration, and so on.

The conceptions of the middle classes offered by Marx and Weber are of extreme importance for our understanding of their overall theories and, as we shall discuss, are still central to present debates over the nature of classes in modern societies. If there is one fundamental difference between Marx and Weber on the subject of class structure, it is their treatment of the middle class—more specifically, the salaried nonmanual segment. We should stress, first, that the two writers largely agree on the disposition of the petty-bourgeoisie portion of the middle classes. Both writers note the existence of this class in early capitalism but foresee its gradual reduction as capitalism develops. The reader will recall that Marx views the petty bourgeoisie as a group that would gradually be swallowed up in the growing concentration of productive property in the hands of large-scale capitalists. Similarly, Weber contends that the chance for individuals to become "self-employed small businessmen" has become "less and less feasible" (Weber, 1922: 305).

On the question of the salaried nonmanual class, however, Marx and Weber take quite different positions. Despite Marx's emphasis on a two-class model of the capitalist class structure, he too is well aware of this salaried nonmanual group, of "the constantly growing number of the middle classes, those who stand between the workman on the one hand and the capitalist and landlord on the other" (Marx, 1862: 573). Given his main purposes, however, Marx does not deal with this class category extensively, for he does not believe that it will be a force of consequence in the ultimate shift from capitalism to socialism. To Marx, the salaried middle class is essentially just an auxiliary to the class of owners, a set of nonproduction employees who, because of the great surplus wealth generated in the more advanced stages of capitalism, can be hired to perform clerical, technical, and minor administrative services for the bourgeoisie (Marx, 1862: 571). Their labour may indeed provide salaried middle-class employees with more of the distributed economic rewards than the industrial proletariat. However, this is an unjust advantage that they have over regular production workers, for they are yet another "burden weighing heavily on the working base," and their collaboration with the bourgeoisie increases "the social security and power of the upper ten thousand" (Marx, 1862: 573). Still, they are in key respects akin to the working class, being in a relationship of exploitation by the bourgeoisie and dependent on the capitalist for wages. Besides, according to Marx, they too will become proletarians over time, as the capitalist system eventually converts even "the physician, the lawyer, the priest, the poet, the man of science, into its paid wage labourers" (Marx and Engels, 1848: 62).

The Weberian conception of the salaried middle class differs from Marx's view, in that Weber often stresses the distributive inequalities that Marx downplays. Members of this class have better economic life chances and different economic interests, which are in themselves sufficient to identify them as a class that is distinct from the workers and that is unlikely to take part in fomenting socialist revolution. The salaried middle class, in Weber's view, will continue to expand in numbers and importance as the children of workers and of the petty bourgeoisie move into the market for white-collar posts in the growing bureaucratic organizations of modern society (Weber, 1922: 305). As well, some members of the non-manual middle class, in addition to their distributive advantage over the ordinary workers, occupy jobs in middle management that place them in relations of domination over other workers. This placement of salaried workers contributes, in Weber's conception, to a clear split between the middle and working classes and sharply reduces the chances that salaried employees will support or identify with the working class in revolutionary action.

Multiple Power Bases: Class, Status, and Party

It is clear from our previous review of Weber's conceptions of class and social class that he departs from Marx primarily in his delineation of several, not just two, important class segments. This view of the class structure is but one element in Weber's more general treatment of social inequality as a pluralist phenomenon. Hence, the analysis of class inequality, in Weber's mind, must be interwoven with, and compounded by, the examination of two additional ideas that are conceptually different from class but that can cut across class distinctions in real societies. The two concepts are *status* and *party*.

As a starting point, we should note that Weber's famous treatment of class, status, and party is only part of his larger essay on "political communities" (Weber, 1922: chap. 9). Thus, underlying the important differences among these three concepts, which we will discuss in the next two subsections, is their common root in Weber's political analysis. A key concern in all of Weber's work is with the "politics" of social life, which broadly speaking is essentially a "struggle" or "conflict" among individuals or groups with opposing interests and different power resources (Weber, 1922: 1398, 1414). *Power* here is the chance (or probability) that individuals or groups can exercise their own wills, even in the face of resistance by others (Weber, 1922: 53, 926). More will be said on the subjects of power and politics in subsequent sections of this chapter. For now, it is important to be aware of the pivotal roles these two ideas play in Weber's sociology and to recognize that class, status, and party are all aspects of "the distribution of power within the political community" (Weber, 1922: 926). Each one represents a potential basis for coalition or organization in the pursuit of personal interests.

Classes versus Status Groups

We have already seen that Weber's classes are categories of individuals differing in economic clout, in "the power ... to dispose of goods or skills ... in a given economic order" (Weber in Gerth and Mills, 1967: 181). *Status*, however, is something that normally inheres in real groups, not simple categories of individuals. In fact, Weber uses the German term, *Stände,* in this part of his analysis, which is typically translated as "status groups," although some theorists prefer to call such entities "estates" (for debate, see Dahrendorf, 1959: 6–7; Bendix, 1962: 85; Wenger, 1980, 1987; Scott, 1996). In either case, however, Weber is clearly speaking of sets of people that are not mere aggregates of individuals but have a subjective sense of common membership and a group awareness that is relatively well defined (Weber, 1922: 932). In addition, status groups tend to have a distinctive "style of life," or mode of conduct, that separates them from the rest of the population (Weber, 1922: 305, 932). Whereas class membership denotes the extent of one's power in the economic order, status groups are delineated by the power that derives from the "social honour, or prestige" distributed within the "status order" (Weber, 1922: 926–927). Thus, the economic-class order and the status order are two distinct hierarchies for representing the relative powers of individuals and groups.

Having drawn this distinction, Weber then proceeds to complicate his analysis by indicating conceptual overlaps between class and status. One crucial overlap to clarify is the connection between Weber's idea of status group and his concept of social class. It will be recalled that Weber characterizes social classes as economic classes whose members, to varying degrees, have acquired some sense of group consciousness and a subjective awareness of their common class position. But, as we have noted, subjective awareness is also a basic element in what Weber calls a status group. The key difference here is that status groups are distinguishable from one another not on economic grounds but in terms of social honour. However, whenever economic power is also a basis for both social honour and subjective group awareness, then a social class is also a status group. Thus, Weber asserts that "the status group comes closest to the social class" (Weber, 1922: 305). Stripped of its subtler details, what this amounts to is a simple equation: when an economic class, a simple category of people with similar economic power, also takes on the subjective awareness and cohesion of a status group, the result is a *social* class.

It is clear from this interrelation between social class and status group that Weber does not seek to portray the economic and status orders as unrelated in actual societies. In reality, the same people can be, and frequently are, ranked similarly in both orders. Thus, for example, those of the highest economic class also tend to possess high levels of status honour. Property ownership in particular is linked to status honour "with extraordinary regularity" (Weber, 1922: 932). Weber asserts that in modern times a person's class location has increasingly become "by far the predominant factor" (though not the only factor) in determining an individual's status group affiliations (Weber, 1922: 935). The subjective awareness basic to status-group membership typically leads to efforts at closure, or the exclusion of outsiders from interaction with members. This interaction can include everything from social activities and gatherings to marriages within the status circle. Here, too, Weber draws some of his examples of status groups from the economic order. Thus, he suggests, at least partly seriously, that because of group closure "it may be that only the families coming under approximately the same tax class will dance with one another" (Weber, 1922: 932).

The purpose, then, of drawing a distinction between classes and status groups is not to assert their total disjuncture but to indicate that they are not necessarily related in a perfect one-to-one correspondence. The power they bestow comes from somewhat different sources, and the extent of their correlation is a matter of empirical investigation in real societies (but see Wenger, 1980). The point is that as long as it is possible for individuals to derive social honour from noneconomic considerations, then the existence of status groups will "hinder the strict carrying through of the sheer market principle" (Weber, 1922: 930). As an illustration, the leaders of the Roman Catholic Church—the Pope and the College of Cardinals—might be said to approximate a status group, to the extent that their power, or ability to influence the actions of others, does not stem primarily from their market position or economic class, which would be below that of the wealthiest capitalists, for example. Instead, the power of church leaders to exercise their will over others is accorded them largely because many people believe them to be worthy of obedience, as God's representatives on earth.

It is important to note as well that, at any one time, a person will belong to only one class but may belong to (or be excluded from) many different kinds of status groups. In other words, there are numerous potential bases for status group formation in society. In fact, Weber argues that "any quality shared by a plurality" can be the basis for being included in or excluded from a status group (Weber, 1922: 932–933). Some of his examples include religious affiliation, ethnic origin, and place of residence (Weber, 1922: 933–936). Elsewhere, Weber also speaks of similar distinctions among "interest groups." In these situations, one group takes "some externally identifiable characteristic of another group—race, language, religion, local or social origin, descent, residence, etc." as a pretext for excluding others (Weber, 1922: 342).

Such exclusion is typically associated with important group differences, not just in people's sense of prestige or social standing, but also in the resources and privileges they possess. A modern illustration involving race is the relationship between whites and nonwhites under the old *apartheid* system in South Africa. In that society, whites could be said to form a status group relative to nonwhites, in the sense that all whites, regardless of their economic or class location, tended to enjoy comparatively greater social standing or prestige, as well as special privileges and prerogatives, simply by virtue of their skin colour. Another example involves gender and the relationship between males and females. In most societies, especially in *patriarchal* systems from past and present, men could be viewed as a status group, in that men generally are accorded more prestige, power, and privilege than females, largely on the basis of a mere physical distinction.

These are just some illustrations of how considerations of status can cut across and complicate considerations of class. Such examples neither deny the great power of those who control the economy nor disprove the possible empirical connections between class position and status group membership. But, for Weber, such examples do reveal that the correlation between the economic order and the status order is not perfect and that, as a consequence, the complexity of social inequality must be acknowledged.

Party

The third concept in Weber's three-pronged treatment of the distribution of power in society is his idea of *party*. Typically, this concept receives far less attention than either class or status. This is probably because Weber himself offers only a brief discussion of party. Some confusion over the significance of party in Weber's scheme has been engendered by the incomplete nature of his discussion. A related difficulty is that Weber identifies power as the specific concern of parties, leading some writers to conclude, erroneously, that *only* parties are concerned with power and that classes and status groups deal with quite different matters. In fact, as has already been noted, Weber argues that all three concepts pertain to the distribution of power (see Wenger, 1987: 44). Each of the three concepts represents a different basis through which power inheres in some groups or individuals more than others.

Parties, in Weber's terminology, are voluntary "associations," systematically organized for the collective "pursuit of interests" (Weber, 1922: 284–285). The best-known examples are formal political parties, such as the New Democratic Party in Canada or the Republican Party in the United States. However, Weber's definition of a party includes a variety of other organizations as well. Some parties may come together to promote special class-related interests; this would be true, for example, in the case of workers' unions such as the United Auto Workers, or middle-class professional associations like the American Medical Association, or pro-capitalist organizations like Canada's Canadian Council of Chief Executives. Other parties may coalesce around noneconomic or status group concerns; some examples include the National Organization for Women, the Assembly of First Nations, or the American Indian Movement. Still other parties may be pressure groups or special-interest organizations, such as Planned Parenthood, Greenpeace, or the National Rifle Association.

Like the members of status groups and social classes, members of parties possess some sense of group consciousness and solidarity. In fact, as some of our examples suggest, status groups or social classes can also be parties in certain circumstances, provided they develop a rational structure, formal organization, and administrative staff. These organizational qualities are primarily what separate party from the other two concepts (Weber, 1922: 285, 938). Thus, not all classes or status groups will be parties, and not all parties will be social classes or status groups, but they can and will overlap in some cases. The message Weber again conveys here is the pluralist nature of both inequality and the structure of power. Parties, like classes and status groups, are distinct features in the intricate mix of social forces that operate in and shape social structures. In contrast to those Marxists who stress economic power as the single force of consequence in the study of inequality, Weber perceives three major bases for power, according to which different constellations of interests emerge. In combination, classes, status groups, and parties all help to determine the nature and extent of social inequality. The concepts of status group and party, moreover, can be taken to subsume a wide array of potentially competing or conflicting social factions, only some of which are constituted around economic issues.

Power, Domination, and Legitimate Domination (Authority)

The complexity of Weber's image of inequality becomes more elaborate still when we delve into his analysis of power in social life. Power is one of the more difficult ideas to work with and comprehend in Weber's writings. This difficulty is both unfortunate and somewhat ironic, because power is also the central concept in much of his sociological analysis, particularly that which deals with social inequality. His general view is that inequalities between social actors are traceable primarily to their differential success in the continuing social struggle, the contest among competing or conflicting interests. This struggle is also the essence of what Weber means by politics, in the broadest sense of the term. Essentially, power is the factor that determines the outcome of the social struggle and, hence, the nature and extent of inequality. Power (*Macht*, in German), as touched on previously, is defined as "the probability that one actor within a social relationship will be in a position to carry out his own will despite resistance" (Weber, 1922: 53).

The problem with Weber's definition of power is that it is too broad to be very useful. Weber himself concedes that his definition is "amorphous," since "all conceivable qualities of a person" could put that person in a position of power over others (Weber, 1922: 53). There also seems to be an allowance here that power relations can be impermanent and sporadic, shifting dramatically on rather short notice.

A key way in which Weber specifies his discussion is to suggest another idea, *domination,* as a "special case of power" (Weber, 1922: 53, 941). Domination (*Herrschaft,* in German) refers to those power relations in which *regular patterns* of inequality are established, whereby the subordinate group (or individual) *accepts* that position in a sustained arrangement, obeying the commands of the dominant group (or individual). In fact, one could claim that it is just such continuing power arrangements that are most relevant to our analysis of social inequality as a *structured* phenomenon. What the distinction between power and domination allows us to do, then, is to set aside those power relations that are only transient, temporary, or incidental. These situations after all tell us little about the more general, established, and patterned systems of domination that provide most of the framework for inequality in social interaction.

It is important to remember, though, that not all analysts adopt Weber's distinction between power and domination, either in their scholarly writings or in everyday conversation. Consequently, the term power is frequently employed when discussing situations that, in Weber's terms, should more precisely be considered examples of domination. For practical purposes in our analysis, the two words may be seen as more or less equivalent, since domination, as defined by Weber, is the key brand of power that students of structured social inequality typically consider.

We have now outlined Weber's distinction between power in general and the more specific variety of power that we are concerned with, domination. In an earlier section of this chapter we discussed Weber's contention that classes, status groups, and parties are the principal bases for exercising power or domination in society. In other words, those who are members of dominant classes, status groups, and party associations are able on the whole to exact compliance to their wills, on a regular basis, from the remaining population. To say this, however, is not to provide the reasons *why* subordinate factions in these spheres accept or endure their subordination.

There are many possible reasons for why people comply with systems of domination. In Weber's terms, some of these reasons are based on *legitimacy,* while others are not (Weber, 1922: 904). There are essentially three pure types of legitimate domination, or *authority* (*legitime Herrschaft*), that are of note. Subordinates may comply because they perceive special charismatic qualities of leadership in those who rule them (*charismatic authority*), because they genuinely believe both in the legality of their own subordinate position and in the legal right of those in power to be there (*legal authority*), or because they voluntarily accept the traditional right of certain groups to lead them, even if this right is not established in formal law (*traditional authority*) (Weber, 1922: 215).

In addition, however, there are numerous other reasons for obedience that are not truly legitimate. In some cases, domination occurs, not as a result of conscious acceptance of the situation due to charismatic, legal, or traditional factors, but because, out of unthinking custom, habit, or convention, subjects have always been subordinate and entertain no possibility of change; in other cases, subordinates may believe that the power imposed on them is clearly illegitimate, but accept their position because of opportunities for personal advantage or self-interest, fear of the use of physical force or other reprisals, an absence of realistic alternatives, and so on (Weber, 1922: 33–36, 753–758, 946–953). In all of these cases, there is no subjective acknowledgment by subordinates that those who dominate them have a legitimate right to do so.

The question of whether any of the types of domination, legitimate or otherwise, is more prevalent than the others is, as usual for Weber, a matter of empirical investigation. His view is that the predominant type of domination will vary over time and across different places and social situations. A crucial underlying reason for compliance in virtually all cases of domination, and a sanction that can never be discounted, is the possible application of physical force by the dominant group. The threat of physical force is more prevalent and more blatant, of course, in early, primitive societies, where "violent social action" is "absolutely primordial" (Weber, 1922: 904). As societies have evolved, the actual use of physical force has typically decreased. Interestingly, Marx generally agrees with Weber that direct force tends to be used "only exceptionally" in modern societies (Marx, 1867: 737; see Sayer, 1991: 68). This reduction in violence is partly because ruling groups are reluctant to resort constantly to violent means, since these actions generate resentment and potential rebellion among subordinates. In addition, the application of coercive force is less necessary, with periodic exceptions, because the other reasons for compliance gradually come into play.

It is usually a combination of some or all of the reasons listed earlier that promotes acceptance (Weber, 1922: 263). In many cases, the rational calculation of self-interest is quite important, especially in the pursuit of economic rewards and social honour (Giddens, 1971: 156). As well, the fact that a "legal order" comes into existence is a significant aspect of domination in modern times. It entrenches in law the rights of certain groups relative to others and so can be used by the ruling group both to justify compliance among the willing and to enforce obedience among the unwilling (Weber, 1922: 312, 903–904). However, there is always the impact of habit and custom to consider, the social inertia that can set in because of the "unreflective habituation to a regularity of life" (Weber, 1922: 312). This can be a particularly important and frequently overlooked factor that contributes to the passive acceptance, by the mass of the population, of domination in everyday life. We shall deal again with many of these ideas about power in subsequent chapters.

Weber's outline of the several reasons that may motivate social actors to accept domination is clearly consistent with his stress on the importance of the subjective meanings that people attach to their behaviour. By elaborating the various types of domination, legitimate and otherwise, that can operate in society, Weber is also providing still more evidence of the complex nature of social processes and the need to take this complexity into account in social explanations. His formulations of power and domination make it clear, as well, that the inequalities among individuals or groups arising out of the economic sphere, or from the class structure, are but one means by which power is exercised and one means by which some people are able to dominate others. This last point is worth stressing because, as we shall see, it has been used by some more recent theorists to suggest that power is the pivotal concept for understanding a full range of other inequalities among groups, including those that are differentiated by race, gender, ethnicity, and age, among others. Although Weber's explicit discussions of these specific concerns are relatively brief (but see Collins, 1986a: 15–16), his recognition that there are multiple nonclass bases for power and group formation can be extrapolated to explain or accommodate these other forms of social inequality.

Rationalization, Bureaucracy, and the Means of Administration

In our analysis of Marx, we noted that his view of social inequality identifies the dominant people in society as those who control the means of material production. To Marx, the explanation for how systems of inequality arise and change is ultimately rooted in how classes emerge out of the organization of economic activity. It is more difficult to identify in

Weber's work the same sense of a single, overriding force shaping social inequality. In fact, the central theme of our discussion of Weber has been to point out his complex view of social inequality and his skepticism about the existence of only one ultimate cause.

Nevertheless, there is some basis for the claim that Weber stresses a recurring, if not a singular, process at work in the generation of social inequality. This claim relates to our earlier observation that Weber sees social action generally as a *contested* activity, a competition or struggle among individuals or groups pursuing their own interests. Factions differ in their power to achieve their interests, and social inequality, in all its guises, is one major result of this imbalance. The contest or struggle for power, which is the crux of what Weber means by politics, is inherent in social action. Thus, if there is a predominant tendency in Weber's approach, it is to explain social action in terms of politics, defined in this broad sense (Giddens, 1972: 34; 1973: 46–47). Such a perspective differs from Marx's, since Weber views the struggle among economic classes as just one element, albeit a crucial one, in the more universal contest of interests within social structures.

A full appreciation of Weber's power and politics orientation to social inequality requires that we consider three closely connected concepts in his work. These are rationalization, bureaucracy, and the means of administration.

Rationalization

Weber perceives in the historical development of societies a tendency for social action to become increasingly rationalized—that is, to be guided by the reasoned, calculated, or rational pursuit of particular interests. Of course, nonrational actions—those prompted by habit or emotion, for example—do continue to occur in modern societies (Weber, 1922: 26). Nevertheless, there is in Weber's view a steady trend toward a predominance of both rationally motivated action and, consequently, rationally organized social structures. This is most obvious in the economic sphere under capitalism, where the calculated pursuit of self-interest reaches an advanced stage (Weber, 1922: 71). Economic enterprise becomes geared above all to the efficiency of production and its enhancement through the systematic organization of business activities (Weber, 1905: 76). The calculation of profits becomes much more precise with the advent of a money-based economy, formal accounting procedures, bookkeeping, and the retention of records and files (Weber, 1922: 107–108). The growth in the specialization of occupations in the workplace and the planned division of labour in large-scale factories and other enterprises are also important features of this economic rationalization (Weber, 1922: 436, 1155–1156).

As we have seen, Marx is well aware of these processes in capitalist development. However, he believes that the growing division of labour and related social structural changes will provide the ultimate impetus toward class polarization and eventual revolution, whereas Weber sees economic rationalization as but one part of a universal trend toward the systematic organization and administration of virtually all social action. Weber notes that a similar process of rationalization has occurred in every major organizational sphere, not just the economic realm. In religion, for example, he sees a trend toward *secularization*, whereby religious practice becomes structured, routinized, and administered much like the activities of other organizations, with a formal church hierarchy, standardized rituals, and so forth. This process is accompanied by a decline in the significance of the magical and the mystical aspects of spiritual life. The change in religion is perhaps the best illustration of Weber's idea that there is a progressive "disenchantment of the world" (Weber, 1922: 538; see also Gerth and Mills, 1967: 51).

In the same way, Weber points out the development of a rationalized legal order and political system. In early societies, a legal system emerges as a basis for legitimizing the monopoly of physical force by those who control a territory. As this set of laws becomes increasingly elaborate and complex, it eventually includes rules and regulations for a wide range of social action. Thus, the original ruling group gradually takes on a variety of adminis- trative roles, leading to the establishment of the modern *state*. The functions of the state are numerous: the enactment of new laws; the preservation of "public order"; the protection of "vested rights" such as property owner- ship; the administration of health, education, and social welfare; and the defence of the territory against outside attack (Weber, 1922: 655, 905, 908– 909). Each of these activities in any large-scale society requires a per- manent, organized system for solving problems, making decisions, and implementing policies. In this manner, the trend to rationalization culmi- nates in the system of formal organizations known as bureaucracies.

Bureaucracy

To Weber, rationalized bureaucracies have become key players in the general power struggle. Bureaucracies of various types have existed from early times. However, the modern bureaucracy is distinguishable from all previous forms by a particular set of traits. We need not examine these characteristics in detail, but they include the existence of specialized occupations, or offices, with designated duties to perform, arranged in a hierarchy of authority or decision-making power. Management of the organization is based on written documents, or files, and the operation is conducted according to an explicit set of rules or administrative regula- tions (Weber, 1922: 956–958).

In Weber's view, such an organization acts almost like a "machine," providing the most rational and efficient means yet devised for adminis- tering social activity (Weber, 1922: 973). This claim of rational efficiency may seem strange to those who have been caught up in bureaucratic red tape and who thus equate the term bureaucracy with irrationality and inefficiency. Certainly, Weber himself speaks of the "impediments" that the bureaucratic apparatus can create in individual cases (Weber, 1922: 975). Nevertheless, he firmly contends that the administration of modern societies would be much worse, in fact impossible, without the bureaucratic form, given "the increasing complexity of civilization" (Weber, 1922: 972). To Weber, this is not a question of preference or a value judgment but a statement of fact. The modern system of rationalized bureaucracies is for Weber a necessary evil: necessary because it is the only practical means for organizing human conduct in the present day, yet evil because it is an "iron cage" that restricts individualism and threatens both democracy and freedom (Weber, 1905: 181–183). We shall return to this point in the concluding section of this chapter.

The Means of Administration

The significance of rationalized bureaucracies for the analysis of social inequality is that they become the key players in the overall power struggle. Bureaucracies are the best examples of what Weber means by enduring structures of domination. They provide the means by which social action is governed on a regular basis and through which a system of inequality is established and sustained.

Once again the contrast with the Marxist conception is noteworthy. Whereas Marxism sees control of the means of economic production as the foundation for class structure and inequality in capitalism, Weber contends that differential access to the means of production is just one of the various ways in which power differences and inequality arise. Each of the major institutional structures of capitalist society has its own sphere of influence and its own bureaucratic system of operation. It is the control of all these various *means of administration* that determines social inequality, not just control in the economic sphere. Hence, in addition to the power deriving from control of the economy by bureaucratically organized corporate enterprises, power inheres in those groups that administer the religious system, the communications media, and so forth (Weber, 1922: 223–224).

Among the most crucial structures for Weber is the interrelated set of "public organizations" that make up the modern state: the legal structure, the judicial and executive branches of government, the civil-service bureaucracies, the police, and the military (Weber, 1922: 980–989). All of these administrative systems hold power in their own right, so that the exclusive emphasis on the economic structure by some Marxists leads to distortion and oversimplification, in Weber's view. Here Weber clearly rejects the notion that the organizations existing outside the economic sphere are merely superstructural props for a ruling class of capitalists.

While acknowledging the great power wielded by the bourgeoisie, Weber appears more worried by the threat posed by bureaucratic officials, especially those found in the state, where the means of administration seem to be increasingly concentrated. The modern state has also shown its growing significance on the international scene, particularly because of its formal power to conduct and control both political and military initiatives in foreign lands (Weber, 1922: 910–925; see also Collins, 1986a: 145–146). It is not surprising that, as we shall discuss later, the power of the state has become a much more prominent concern for both non-Marxist and Marxist scholars, since Weber's time.

The Future: Democracy, Bureaucracy, and Socialism

Our outline of Weber's work has revealed his complex, multifaceted conception of social structure and the dynamics of social inequality. Modern society is shaped by an ongoing contest or struggle among self-interested social actors. Through control of economic resources, the acquisition of status honour, and the influence deriving from party associations, power is exercised over others. The power struggle gradually is consolidated into a contest involving formal structures of regular, patterned domination. Increasingly, in modern times, these structures take the form of rational bureaucracies, each headed by its own select group of high administrative officials. Social inequality in this context is firmly established, and power is concentrated in the hands of bureaucratic officials in all spheres of influence, not just the economic. A marked concentration of control over the administration of society falls to the various branches of the state.

It should be apparent from this description of Weber's position that he does not share Marx's optimism about the transformation of capitalism into a new form of society that ensures universal human freedom and equality. As has been noted, Weber himself is sympathetic to the cause of political democracy. However, because of the structures of domination that have evolved under capitalism, Weber remains pessimistic about the prospects for democratic action and for the elimination of inequality in future societies. Moreover, for a number of reasons, he is convinced that socialism cannot reduce, and may even intensify, the extent of bureaucratic domination in the future.

Democracy versus Bureaucracy

Democracy, in Weber's definition, is a system of government whose guiding principle is "the 'equal rights' of the governed," a system in which the power of "officialdom" is minimized and the influence of "public opinion" is maximized (Weber, 1922: 985). Such a system is most closely approximated in "direct" or "immediate" democracies—those in which persons in authority are obligated to conform to the will of their constituents (Weber, 1922: 289). The problem with this brand of democracy is that it cannot function in large-scale systems like a modern society because of the sheer size and complexity of such structures (Weber, 1922: 291). The only alternative is to employ what Weber calls "representative democracy," in which constituents elect individuals who are empowered to act in accordance with the general interest. This "representative body" of elected officials in turn typically appoints or elects a subset of their number as a "cabinet," which itself may be overseen by a premier official, or prime minister, for purposes of centralized coordination of action. Such a system of "parliamentary cabinet government" is the most familiar form of modern democracy at the mass level (Weber, 1922: 289–297).

But what, after all, is this system of government? It is clearly another instance of bureaucracy, with the same inherent structure of domination that such a designation implies. The dilemma for establishing and protecting democracy in the present and the future is that "bureaucracy inevitably accompanies modern mass democracy" (Weber, 1922: 983). A bureaucratic form of organization is the only means by which democratic action is feasible in complex societies. It is the one structure that ensures regular, predictable administration while at the same time providing procedures for the protection of constituents against possible abuses of democratic principles by those who govern. That is, governmental powers may themselves be constrained by a democratic system of laws and regulations that includes appeal procedures, explicitly limited terms of office, multiple parties, provision for elections, and so forth.

However, government bureaucracy is also a hierarchy of authority, influence, experience, and expertise. Thus, even in a democracy, government leaders are a select group with "special expert qualifications" and often have access to more information and resources than any other group (Weber, 1922: 985). This means that the very same bureaucratic system that is necessary for democracy in any complex modern society creates conditions that could be the cause of undemocratic action by those who rule. Of course, this is not to say that such abuses of democracy *must* occur. On this point, Weber disagrees with his contemporary, Robert Michels, who asserts that an "iron law of oligarchy" operates in organizations, ensuring that they always abandon the democratic process in the end (Michels, 1915). Weber's position, as always, is more conditional: "democracy inevitably comes into conflict with the bureaucratic tendencies" of modern society (Weber, 1922: 985); however, this only means there is a probability that democracy will be undermined, not that it must fail, with certainty.

This idea comes up again in Weber's observation that political action is always determined by the "principle of small numbers," which gives superior "maneuverability" to political elites or "small leading groups" (Weber, 1922: 1414). He even says in the same passage that "this is the way it should be," but only as long as leaders are both successful and faithful to essential democratic principles (Weber in Gerth and Mills, 1967: 42). For Weber, the main concern in democratic politics is effective leadership (Giddens, 1972: 54). His personal preference is for a responsive political democracy, headed by someone possessing special qualities of democratic enlightenment and personal charisma. His fear, however, is that politics in the future will take the form of a "leaderless" democracy, with the mass of the population controlled by a "clique" of bureaucratic officials, a "certified caste of mandarins" (Weber in Gerth and Mills, 1967: 113, 71).

The Socialist Alternative

Given all of the previous discussion, Weber's view of the prospects for democracy is, on balance, more pessimistic than optimistic. Even his own preferred version of political democracy contains within it the potential loss of individual autonomy and freedom, because bureaucracy is an "escape-proof" feature of every type of modern society (Weber, 1922: 1401). It is notable that Marx also discusses the existence of bureaucracy in modern times; however, he seems on the whole to treat bureaucracy as a bourgeois form of organization that, like the capitalist state and bourgeois society generally, will disappear or be fundamentally transformed with the rise of socialism (Marx, 1843: 45–47; see Sayer, 1991: 78–79). However, to Weber, any claim that socialism is the solution to the problem of bureaucracy, or the means to realize true democracy, is simply mistaken. The imperative need for the systematic organization of social action is universal: "It makes no difference whether the economic system is organized on a capitalistic or socialistic basis" (Weber, 1922: 223). If anything, because socialist systems tend to intensify the centralization of decision making in the state and intrude pervasively into all spheres of human activity, "socialism would, in fact, require a still higher degree of formal bureaucratization than capitalism" (Weber, 1922: 225). In this way, socialism would further reduce, not increase, human freedom. In other words, any hope for a dictatorship of the proletariat is a delusion. Instead, socialism would be an extreme case of "dictatorship of the official" (Weber in Gerth and Mills, 1967: 50; see also Löwith, 1982: 53, 106).

It appears, therefore, that Weber does not share with Marx the expectation of an eventual end to social conflict and to structured inequality. To Weber, the need for bureaucracy and the factionalism of human interaction make both inequality and competitive struggle inherent features of all societies, socialist or capitalist. One of the key differences between these two great thinkers seems to centre on this point. Both give the idea of struggle a key role in their conception of inequality. But for Marx, on the one hand, struggle is not essential to social life. It is part of society only as long as class systems exist and act to pervert the essentially social and cooperative nature of human interaction. For Weber, on the other hand, struggle in some form is basic to all social life and operates generally, not just in the sphere of class relations. In his view, inequality emerges from the continuing contest for power, in which individuals and groups often pursue, not the general interest, but their own special interests. One can seek to place checks and balances against the usurpation of power by bureaucratic officials, but such efforts may not be successful and, in any case, will eliminate neither inequality nor the self-interest inherent in human nature.

Which of these images of our present and future is the more accurate one, of course, remains a topic of debate to this day. Ultimately, it is a question with no simple answer. In assessing the future of inequality, some might deride Marx for the shortcomings of optimism: his failure to discern all the obstacles to revolution; his omission of a clear guide for his disciples to the universal society; and his perhaps excessive faith in humanity. Still, others have argued that Weber's hard-nosed pessimism is more flawed (Marcuse, 1971). Some might contend that, in predicting for us a future of bureaucratic cages and embattled democracy, Weber has inadvertently helped contribute, in a self-fulfilling way, to the realization of his own worst fears.

SUMMARY

Our concern in this chapter has been to outline and assess Max Weber's conception of social inequality. Weber's work may be characterized generally as a positive critique of Marx and Marxism. A central theme in Weber's writings is his emphasis on the complex, pluralist nature of inequality. This theme was illustrated by reference to a number of Weber's formulations, especially his conceptions of class, power, and domination. We noted Weber's view of society as an arena for numerous contests among social actors attempting to obtain and exert power, and we examined the important role played by bureaucratic structures in these struggles. Finally, the chapter concluded with Weber's assessment of the chances for democracy in future societies, both socialist and capitalist.

CHAPTER FOUR

Durkheim, Social Solidarity, and Social Inequality

> *Liberty ... is itself the product of regulation ... Only rules can prevent abuses of power.*
> ÉMILE DURKHEIM
> *The Division of Labor in Society*, Preface to the second edition, 1902

INTRODUCTION

In this chapter, we consider the work of Émile Durkheim. Durkheim is generally thought to be the third major figure in classical sociological theory. His ideas, along with those of Marx and Weber, are still believed by various writers to lie at the heart of most theoretical developments and debates in sociology today (Coser, 1977: 174; Wiley, 1987: 25; see also Collins, 1985; Sydie, 1987). However, although Durkheim's contributions to general sociological theory have received wide recognition, the value and relevance of his ideas for the analysis of social inequality have not been widely acknowledged (for some exceptions, see Grusky and Sorensen, 1998; Grusky, 2005). This relative lack of acknowledgment is partly to be expected, since Durkheim's writings are not primarily concerned with the problem of inequality. Nevertheless, it is also the case that Durkheim has a good deal more to say about inequality than is normally assumed. Moreover, his conceptions share some interesting affinities with the writings of both Marx and Weber. Finally, Durkheim's work serves as a useful bridge between the classical theories of inequality and the more recent structural-functionalist perspective on social stratification, which became prominent after Durkheim's era and which is our major concern in Chapter 5. All of these considerations indicate the importance of examining the ideas put forth by Durkheim on the problem of social inequality.

It is somewhat surprising that, despite being almost exact contemporaries, Durkheim and Weber seem not to have been interested in or influenced by each other's work (Lukes, 1973: 397; Parkin, 1992: 4). Durkheim was aware of Weber's writings and occasionally made reference to some of them (Giddens, 1971: 119). Even so, Weber stands out as the only major figure of the period whose ideas escaped any substantial critique or evaluation by Durkheim (Lukes, 1973: 405). Durkheim did indicate more extensive interest in the theories of Marx (see Alexander, 1990a: 126–127, 149). His thought even reveals some broad similarities with Marx's work, partly because of the influence of Saint-Simon on both writers. Nevertheless, the affinities between Marx and Durkheim are rarely direct or explicit ones, as evidenced by the relatively few and largely incidental citations of Marx in Durkheim's writings (e.g., Durkheim, 1893: 393; 1896: 12, 14, 59, 221; see Giddens, 1971: 97).

This shortage of direct links between Durkheim and either Marx or Weber is to some extent the result of Durkheim's specific sociological interests—interests that are largely distinct from those of the other two writers. It is true that the three men share a concern with analyzing and comprehending certain aspects of the origin and development of nineteenth-century capitalism. However, each writer concentrates on a particular feature of this societal form. Marx focuses on class struggle—the opposition between owners and nonowners in the productive sphere—and its role in the emergence, maturation, and anticipated demise of capitalism. In contrast, Weber stresses the more general power struggle, especially with regard to the growth of rationalized bureaucracies, which become the principal structures of domination in modern capitalism but which, for Weber, are inescapable in socialism as well.

To Durkheim, also, problems of class struggle and power inequities in capitalism are important, but he analyzes them primarily with reference to what for him is the prior and more crucial question: how are the social structures within which power and class struggle operate even possible in the first place? By what means does a scattered collection of people coalesce to form the ongoing, enduring entity we call society, when so many conflicting and divisive forces are capable of tearing it asunder? Durkheim's focus, then, is on the cohesiveness, or *solidarity*, of social arrangements.

Durkheim's concern with the question of social solidarity, as well as his general perspective on social inequality, can be more easily comprehended if placed in the historical and intellectual context of his time. As with Marx and Weber, then, our assessment of Durkheim begins with an account of the major life events and other experiences that helped to shape his thought.

BIOGRAPHICAL AND INTELLECTUAL SKETCH

Early Years

Émile Durkheim was born on April 15, 1858, at Épinal, in the Lorraine region of eastern France. His father, grandfather, and great-grandfather had all served as rabbis in Lorraine and the nearby province of Alsace. These two areas were home to the largest Jewish community in France at the time (Peyre, 1960: 7). This community had been in existence for several centuries but became increasingly well established with the granting of full civil rights to Jews after the French Revolution (Lukes, 1973: 39; Coser, 1977: 161–162). Durkheim was confirmed into the Jewish faith at age thirteen and showed some initial interest in being a rabbi like his father; however, he soon changed direction and eventually turned away from all religious involvement (Coser, 1977: 143). Nevertheless, like Weber, Durkheim remained a "non-believer fascinated by religion" (Parkin, 1992: 6). His interest in religion and its role as a source of social cohesion in both early and modern societies had a lasting impression on much of his sociological thought.

Durkheim appears to have enjoyed a close-knit and supportive family life, but one in which the values of hard work and material austerity were strongly emphasized (Lukes, 1973: 39–40). Even in this early period, then, there is evidence that Durkheim developed a sense of duty, diligence, and discipline that was to mark virtually all the other stages of his life (e.g., Coser, 1977: 144).

Probably the most serious historical event in Durkheim's early years was the defeat of France by Germany in the brief Franco-Prussian War of 1870–71. As a young boy, Durkheim observed some of the grave consequences of this defeat first-hand, including the flaring of anti-Semitism and the tendency by some to blame members of France's Jewish population for the loss of the war. This attitude must have been both hurtful and inexplicable to Durkheim and many other Jews in Alsace-Lorraine, since their community was known for its patriotism and for providing France with many of its career army officers and civil servants (Lukes, 1973: 39–41). In spite of such prejudice, however, Durkheim never abandoned his loyalty to his country. He also remained deeply committed to the central ideals of the French Revolution, including the belief in individual liberty and progressive democracy (Lukes, 1973: 47, 271, 546; Giddens, 1986: 13–15). This point is important to emphasize, since, as Lukes has shown, Durkheim at times has been incorrectly characterized as a reactionary opponent of political liberalism and personal freedom (Lukes, 1973: 2–3, 338). On the contrary, Durkheim's contention is that liberty is an essential good in modern societies, provided that it is tempered by a sense of duty and moral obligation to others (Peyre, 1960: 21).

Durkheim as a Young Scholar

From the beginning of his scholastic career, Durkheim was recognized as an outstanding student. As a young schoolboy, he was judged to be sufficiently advanced that he skipped two full classes beyond his own age group (Lukes, 1973: 41). The one period of academic adversity that Durkheim faced was during his late adolescence, when he sought admission to the prestigious École Normale Supérieure in Paris. Accounts suggest that a variety of circumstances, including financial problems, family illness, and the failure of his examiners to appreciate his original mind, contributed to a three-year delay before he was finally accepted into the École in 1879 (Peyre, 1960: 10; Lukes, 1973: 42). From that point on, however, Durkheim enjoyed virtually total success in his scholastic and academic pursuits.

Life at the École was both spartan and strict, but it apparently suited the serious and austere young Durkheim, who was thought to be rather aloof and stuffy by some of his fellow students (Coser, 1977: 144; Parkin, 1992: 3). At the same time, though, the École was an extremely stimulating intellectual environment for Durkheim and rightly deserved its reputation for attracting the brightest and keenest young minds in France (Lukes, 1973: 45). Even among this select company, Durkheim was seen as more mature and more able than his classmates, many of whom held him in awe (Lukes, 1973: 52).

Durkheim completed his course of study at the École in 1882 and, like other future French intellectuals of the period, spent several years as an instructor at various secondary schools (lycées) near Paris (Peyre, 1960: 11). He also devoted part of one academic year (1885–86) to a study for the French education ministry, during which he visited and evaluated several German universities, including the University of Berlin (Coser, 1977: 145; Lukes, 1973: 85; Jones, 1994: 4–5). It is interesting that Max Weber, at this same time, was attending Berlin as a student in law. A second parallel of note is that Weber had just returned from his compulsory military service in Alsace-Lorraine, the area of France taken by Germany after the Franco-Prussian conflict and the region where Durkheim was born and his family still lived. Hence, while there appears to be no record that the two men ever knowingly crossed paths, these coincidences make the chance of an unwitting encounter between Durkheim and Weber an intriguing possibility.

In the years that Durkheim served within the lycées, he established a reputation as both an exceptional teacher and a promising scholar. His accomplishments ultimately led to his appointment to the University of Bordeaux in 1887, as head of both Social Science and Pedagogy (Education). This post was created especially for Durkheim, who was seen by the education ministry as someone who could create a French social science to rival that which was already flourishing in the German university system (Lukes, 1973: 95). His joint involvements, in both social science and pedagogy at Bordeaux, in a sense foretold two of his ultimate contributions to French intellectual life: first, as the pioneering figure in the establishment of sociology as a recognized and legitimate field of study and, second, as an active participant in the movement to reshape and modernize the overall system of French education.

The Bordeaux Period

Durkheim's time at Bordeaux was an unusually rewarding period, both personally and professionally. He married and settled down to what has been portrayed as a very happy and secure family life. His wife was an important source of support for Durkheim, as a homemaker and mother to their two children, and as a collaborator and editor in some of Durkheim's research and writing (Lukes, 1973: 99). In the fifteen years that he spent at Bordeaux, Durkheim also produced many of his most famous and important works. These include *The Division of Labor in Society*, which was the subject of his doctoral thesis, as well as *The Rules of Sociological Method* and his sociological analysis of *Suicide*. At this time, Durkheim also established the *Année Sociologique*, an annual review of French sociology that soon became the most influential publication of its kind in the country.

Although Durkheim is sometimes viewed as a remote academic, isolated from the conflicts and contentious issues of his day, there is some notable evidence to the contrary. First of all, beginning with his years at Bordeaux, Durkheim showed in his published writings a clear willingness to engage in heated debate or polemical exchange with other leading intellectuals. Many of his contemporaries disagreed fundamentally with his theories and with his promotion of sociology as the discipline best suited to analyzing such diverse concerns as religion, education, suicide, and crime (Lukes, 1973: 296, 307, 318–319, 359). Second, although it is true that the pursuit of social or political causes was not a major preoccupation for Durkheim, there are several examples of his active involvement in the controversial events of his time. The best illustration from the Bordeaux period was Durkheim's support of the movement to exonerate Alfred Dreyfus, a French army officer of Alsatian Jewish background who had been falsely accused but twice convicted of being a spy for Germany. Durkheim was one of the first academics to speak out and to organize on Dreyfus's behalf; this was characteristic of Durkheim and indicative of his concern with the threat to individual rights and freedoms posed by the case (Lukes, 1973: 333, 343, 347). The Dreyfus affair dragged on for many years and was not settled until his eventual acquittal in 1906. At various points throughout this period, Durkheim was among those who showed a readiness and ability to confront powerful adversaries, including right-wing and often anti-Semitic elements within the military, the judiciary, the intelligentsia, and the Catholic Church (e.g., Parkin, 1992: 5).

Durkheim at the Sorbonne

It was almost inevitable that Durkheim's growing reputation as a scholar led, in 1902, to his being called to Paris and the Sorbonne, France's elite university. Durkheim lived out the rest of his life there, as one of the most eminent and formidable figures in French academia.

Not long after moving to the Sorbonne, Durkheim was named Professor of the Science of Education and, later, by special decree of the ministry of education, was made Chair of the Science of Education and Sociology (Coser, 1977: 147). Just as he had done at Bordeaux, then, Durkheim occupied a dual academic posting that allowed him both to promote sociology as the new science of humanity and to foster important changes in the French education system (Lukes, 1973: 409; Coser, 1977: 148). Durkheim for several years taught compulsory courses in moral education to many of France's future teachers and also introduced sociology into the French school curriculum (Lukes, 1973: 364, 379).

The subject of sociology became increasingly popular and influential in France during the early 1900s, so much so that church leaders and other traditional segments of society feared that the sociological perspective, as presented by Durkheim and his disciples, would come to dominate the minds of an entire generation of French youth. Especially worrisome was Durkheim's secular, nonreligious approach to understanding human values and his apparent preference for the state over the church as the instrument of moral education in modern times (Lukes, 1973: 374–375). Durkheim's contention that the growing division of labour in industrial societies would ultimately replace traditional religious beliefs as the main source of social solidarity was also distressing to more conservative thinkers.

While such concerns for the power of sociology were undoubtedly exaggerated, there is little question that this period was a triumphant one for Durkheim's new science of society. Unfortunately, this lively and burgeoning intellectual climate was tragically and irreparably altered by the outbreak of World War I in 1914. As patriotic as ever, Durkheim immediately turned from his own work to devote most of his energy to the war effort, primarily as an organizer and a researcher for committees publishing studies and other documents on the war (Lukes, 1973: 549). Many of Durkheim's most gifted and promising students, including his son, André, enlisted in the military and paid a heavy price for the French cause. Hundreds would be killed before the war finally ended in 1918, among them over half of one graduating class (Peyre, 1960: 17; Lukes, 1973: 548).

Émile Durkheim,
1858–1917

© Bettmann/CORBIS.

For Durkheim, the greatest loss came in April of 1916, when he learned of his son's death on the Bulgarian front. It appears that this personal tragedy was a blow from which Durkheim never recovered. He took great pains to hide his grief and remain busy with his work (Lukes, 1973: 558). However, his health steadily deteriorated until late in 1916, when he suffered a stroke. After some brief improvement in his condition, Durkheim's heart gave way, and in November 1917 he died, at the age of 59.

In spite of this unfortunate and untimely end, Durkheim's legacy to sociology has been a rich one. In the remainder of this chapter, we will examine some of this wealth of ideas, particularly those that provide insight into the nature of social inequality. Our initial focus will be on Durkheim's conception of social solidarity and its sources in both early and modern societies. We then consider his views on social inequality, with specific reference to his analysis of the division of labour, the role of the contemporary state, and the special problems posed by the economy. In the concluding sections, we consider Durkheim's views on inequality in future societies and compare some of his projections with those of Marx and Weber.

DURKHEIM AND SOCIAL SOLIDARITY

We have already noted that Durkheim's major concern throughout most of his work is with the manner in which societal cohesion or solidarity is made possible. In *The Division of Labor in Society,* Durkheim puts forth his central thesis on social solidarity in human societies. Durkheim's idea of social solidarity suggests an image of society as a set of interconnected groups and individuals, interacting with one another in regular, patterned, and more or less predictable ways. Essential to this interaction is *morality,* or moral regulation, by which Durkheim means the set of rules or norms that guide and govern human conduct. The existence of these guidelines means that individuals interact in accordance with their obligations to others and to society as a whole. In doing so, each person also receives some recognition of his or her own rights and contributions within the collectivity. Morality in this sense is "strictly necessary" for solidarity between people to occur; in fact, without morality "societies cannot exist" (Durkheim, 1893: 51). One of Durkheim's prime concerns is to determine how the forces providing morality, and hence solidarity, have changed as society has moved from early forms to the capitalist system of his time (Giddens, 1971: 106).

One of Durkheim's essential arguments is that social solidarity stems historically from two sources, one of which prevails in early societies and the second of which predominates in the modern era.

Early Societies and Mechanical Solidarity

Early societies are held together by a "mechanical" or automatic solidarity, a union based on the "likeness" or similarity of people (Durkheim, 1893: 70). Such societies are really aggregations of families or other subunits having certain characteristics in common. These subunits have a structural resemblance to one another, being for the most part economically self-sufficient and capable of providing for their own production and consumption needs. Their similarity by itself is not enough to maintain solidarity, since subunits can survive in isolation from one another. In addition, however, early peoples also share a "collective consciousness" (in French, *conscience collective*), a "totality of beliefs and sentiments common to average citizens" (Durkheim, 1893: 79). The moral rules necessary for social solidarity in early times are represented in this body of shared beliefs and sentiments. They provide the social glue that binds together what otherwise would be independent, self-sustaining subgroups. The set of common values is generated within the collectivity and evolves with it over a considerable time, during which adherents presumably come to agree on its central precepts, while dissidents either convert or are excluded.

The collective consciousness, which touches all spheres of life in early societies, finds its principal repository in religion. Early religion embodies not just purely religious beliefs but also a "confused mass" of ideas regarding "law," the "principles of political organization," and even "science" (Durkheim, 1893: 135). The connection between religion and law is particularly significant for Durkheim. In fact, he believes law is "essentially religious in its origin," representing customs and beliefs similar to those in the collective consciousness, but in a more organized and precise fashion (Durkheim, 1893: 65, 92, 110). This link between law and religion is illustrated in the ancient practice of having priests serve as judges and in the medieval concept of the divine right of kings, which granted political and legal powers to monarchs as God's representatives on earth.

In Durkheim's view, then, legal, or *juridical,* rules gradually develop out of the collective consciousness, especially through the influence of religious beliefs. If there is a close correspondence between these legal rules and the collective consciousness, we then have a moral society, one in which individuals acknowledge one another and interact smoothly. However, Durkheim clearly realizes that the formal legal system is not always moral or just. In the course of social change, laws may not change accordingly, making them inappropriate, unjust, and contrary to genuinely moral regulation. Echoing Weber to some extent, Durkheim asserts that such laws may be maintained only through habit, artificial manipulation, or force (Durkheim, 1893: 65, 107–108). Thus, Durkheim recognizes that societies must and will change and that a healthy, or *normal,* society is one in which the regulations that provide the basis for solidarity are suited to the historical period in question.

Modern Societies and Organic Solidarity

The transition of social life from early times to the era of modern capitalism entails fundamental modifications in structural arrangements and, hence, in the form of social solidarity. The old mechanical solidarity, based on likeness and a common set of beliefs and sentiments, is assailed on all sides, most importantly by changes stemming from the expanding division of labour. Here Durkheim reveals that, like Marx and Weber, he perceives the division of labour as a crucial force in the historical evolution of social structures. In contrast to early societies, where subunits resemble one another both in their economic self-sufficiency and in their common belief system, modern societies comprise individuals who tend to be quite dissimilar, engaged in specialized tasks and activities in their daily lives, and guided by distinct norms and values in their personal conduct.

In such a situation, it is difficult to maintain solidarity through the collective consciousness, because a "personal consciousness," with an emphasis on individual distinctiveness, increasingly comes to the fore-front in people's minds (Durkheim, 1893: 167). This is not to say that the collective consciousness disappears altogether; however, the common beliefs it entails gradually become "very general and indeterminate," acting more as maxims or credos than as detailed guides for conduct (Durkheim, 1893: 172). Hence, the collective consciousness provides some integrative force in modern society, but of a vague and general sort. Now "each individual is more and more acquiring his own way of thinking and acting, and submits less completely to the common corporate opinion" (Durkheim, 1893: 137, 152, 172).

The weakening of the collective consciousness is paralleled in the declining influence of religious beliefs. Whereas religion "pervades everything" in early society, in the modern era it "tends to embrace a smaller and smaller portion of social life" (Durkheim, 1893: 169). Now societies are far more specialized, no longer composed of self-sufficient subgroups but rather of subunits that perform specific "functions" that, taken together, contribute to society's existence (Durkheim, 1893: 49). This development of special structures fulfilling particular functions is also evident in the decreasing role of religion: "Little by little, political, economic, and scientific functions free themselves from the religious function, constitute themselves apart, and take on a more and more acknowledged temporal character" (Durkheim, 1893: 169). In addition, of course, tasks within each of these special structures or spheres of activity are themselves divided and specialized, producing the complex network of individual positions and roles that is modern society (Durkheim, 1893: 40).

It is worth noting here that Durkheim does not mourn the passing of early societies or the decline of the collective consciousness. On the contrary, he views the growing division of labour and specialization of functions as normal aspects of modern life. The new emphasis on indi-vidualism and personal distinctiveness in modern societies is also a normal development, something wholly compatible with, even necessary to, the expanding division of labour. Durkheim believes that this "cult" of the individual personality is here to stay and is a positive development, as long as individual goals and interests do not override collective interests and other people's rights, thereby endangering social solidarity (Durkheim, 1893: 172, 400). What is needed is some new integrative force to replace the declining collective consciousness—a force that can "come in to take the place of that which has gone" (Durkheim, 1893: 173). Otherwise, the individualism of modern life will degenerate into blatant self-interest, or *egoism*—a phenomenon that is destructive of both mor-ality and social solidarity (Giddens, 1986: 11, 13).

Durkheim's principal contention in *The Division of Labor in Society*, and one of the most clever twists in his analysis, is that the division of labour is both the main cause for the weakening of the old mechanical solidarity and the key means by which the new form of solidarity emerges. In Durkheim's terms, the function of the division of labour is to provide solidarity on the basis of *dis*similarity, on the fact that highly specialized individuals and subgroups are no longer self-sufficient, but must cooperate and depend on one another for survival (Durkheim, 1893: 228). Durkheim calls this form of solidarity "organic" because it portrays modern society as similar in key ways to an advanced biological organism, "a system of different organs, each of which has a special role, and which are themselves formed of differentiated parts" (Durkheim, 1893: 181, 190–193).

In addition to providing cohesion through this exchange of interdependent services and functions, "the division of labor produces solidarity ... because it creates among men an entire system of rights and duties which link them together in a durable way" (Durkheim, 1893: 406). Increasingly over time, these rights and duties become established in formal laws because their number and complexity necessitate systematic codification. As noted earlier, Durkheim believes formal law originates with religious beliefs and gradually takes over from religion the task of moral regulation as society evolves. This takeover, however, is never complete. Moral rules, in "a multitude of cases," continue to be based on uncodified and nonlegal precepts, on "usage," "custom," or "unwritten rules" that complement and fill in the gaps left by legal rules (Durkheim, 1893: 147, 215).

The nature of these moral and juridical guides for conduct and interaction also changes gradually as society develops. In the past, their prime purpose was to defend and reaffirm the collective consciousness, the common set of beliefs in traditional society. Now, however, they become a complex system of mutual obligations, rights, and duties, which, in ideal circumstances, provide for smooth interaction between individuals, the "pacific and regular concourse of divided functions" (Durkheim, 1893: 406).

DURKHEIM ON SOCIAL INEQUALITY

At this point, the reader may have the impression that Durkheim's portrayal of modern society is an unrealistically positive one. If social structures are moving toward a new organic solidarity in which people cooperate and depend on one another in a network of rights and obligations, how is it that other analysts perceive so many social problems? What of Marx's concern with class conflict and worker oppression, or Weber's fear of injustice and servitude through bureaucratic domination?

Some critics might suspect Durkheim of ignoring these problems or of considering them unimportant. However, a closer reading of Durkheim reveals his awareness of and interest in such issues (see Collins, 1988a). It is true that Durkheim discusses these problems primarily with reference to their effects on social solidarity, which is typically his main concern. Nevertheless, his observations provide evidence that he has a clear conception of the nature and consequences of social inequality, one that shows some surprising parallels with the ideas of both Marx and Weber. We cannot consider Durkheim's views in all their detail, but several key elements should be briefly examined. These include his distinction between the normal division of labour and its *anomic* and *forced* variants; the role of the state as the moral guardian of modern society; the special problem of the economy and the emergence of *occupational groups*; and the prospects for social solidarity and social inequality in future societies.

The Division of Labour

The Normal Form

When Durkheim stresses the positive functions of the division of labour for modern society, he is speaking of a division of labour that approximates the ideal or normal case. The division of labour is normal if there is genuine moral regulation guiding the interactions among people and if there is justice in the attainment of positions within it. The division of labour is *moral* if individuals interrelate with restraint, recognizing one another's obligations and contributions, both to other individuals and to the collectivity as a whole. The division of labour is *just* if each person has an equal opportunity to take on that position most appropriate to his or her capacities and interests. Violation of either of these principles produces an abnormal form of the division of labour, a structure that cannot fulfill the functions Durkheim ascribes to the normal form. Though the distinctions are sometimes blurred, insufficient morality (or amorality) is conducive to an anomic division of labour, whereas insufficient justice (or immorality) tends to create a forced division of labour (Durkheim, 1893: Book 3). Let us examine each of these separately.

The Anomic Form

Whenever the division of labour fails to underscore and reinforce the ties that bind us, moral deregulation, normlessness, or *anomie* can occur. Rather than producing solidarity, the anomic division of labour can promote "contrary results," situations in which mutual contributions and obligations are denied or overlooked, leading to lawless and unconstrained struggles (Durkheim, 1893: 353). A prime example is in the economic sphere whenever capitalists and workers become set against each other, with neither side exercising moderation or compromise. The typical consequences are "industrial and commercial crises," marked by "conflict between labor and capital" such that "a sharp line is drawn between masters and workers" (Durkheim, 1893: 354–355).

Here Durkheim's discussion sounds very much like Marx's analysis of class polarization in capitalism. Like Marx, Durkheim observes the increasing concentration of capitalist ownership in a few hands, with the result that "enterprises have become a great deal more concentrated than numerous" (Durkheim, 1893: 354). This concentration only accentuates the division between owners and workers and makes it even more difficult to build a system based on mutual respect, trust, and moral regulation between classes. The main difference between Marx and Durkheim here is that Marx believes such problems are basic to any capitalist division of labour and any class structure. Durkheim, however, believes that class polarization of this sort occurs because the division of labour in this case is anomic—an abnormal form that requires adjustment.

Before proceeding, we should note another interesting parallel between Marxian theory and Durkheim's anomic division of labour. This parallel concerns the working conditions faced by the proletariat in capitalism. We have seen that Marx perceived the capitalist division of labour as a prime cause of alienation in the working class. In a similar way, using similar imagery, Durkheim describes how the anomic division of labour degrades the individual worker, "making him a machine," "an inert piece of machinery" that performs "routine" tasks "with monotonous regularity," "without being interested in them and without understanding them" (Durkheim, 1893: 371).

The key difference here between Marx and Durkheim is, again, that Marx apparently believes any division of labour has such consequences, whereas Durkheim identifies this "debasement of human nature" only with the abnormal, anomic division of labour, a system that is itself debased and divested of its moral character (Durkheim, 1893: 371). For Durkheim, the solution is not the elimination of the division of labour, which he believes is indispensable to the operation of modern society. Instead, it is essential that all of us perform our roles in the division of labour as best we can, keeping in mind our obligations to our fellow humans and to society. Not all of us will have equally important parts to play, but such equality is not necessary as long as each person has a sense of obligation and contribution, a sense that one's labour tends "towards an end that he conceives" and a feeling that one "is serving something" (Durkheim, 1893: 372). Moreover, it is not essential that each of us have a similarly broad perspective about how things work. A person "need not embrace vast portions of the social horizon; it is sufficient that he perceive enough of it to understand that his actions have an aim beyond themselves" (Durkheim, 1893: 373). These additional elements will alleviate the anomic aspects of the division of labour that have plagued modern capitalism, particularly within the working class.

The Forced Form

The eradication of anomie through moral regulation is for Durkheim basic to any good society. However, regulation alone will not eliminate all of the abnormalities that threaten social solidarity. As Durkheim notes, "it is not sufficient that there be rules," for "sometimes the rules themselves are the cause of evil" (Durkheim, 1893: 374). This point relates to Durkheim's earlier discussion of the possible lack of correspondence between formal laws or rules and true morality. Rules may exist that are no longer appropriate to the time but are retained because they serve the special interests of those in a position to keep them in operation, through force, manipulation, or an appeal to tradition (Durkheim, 1893: 65, 107–108).

It is the application of such inappropriate and unjust rules that characterizes the second abnormal division of labour, which Durkheim calls the *forced* form. In modern society, a key basis for morality and social solidarity is the flourishing of individualism, "the free unfolding of the social force that each carries in himself" (Durkheim, 1893: 377). A normal division of labour is one in which aspects of people's real selves, including their special "aptitudes" and "natural talents," are allowed to develop (Durkheim, 1893: 375). Under such a system, individuals will find their appropriate places in society, thereby satisfying their own desires while at the same time maximizing their ability to fulfill society's needs. Here again we hear some echoes of Marx, who also hoped for a society in which "the free development of each is the condition for the free development of all" (Marx and Engels, 1848: 95). Unfortunately, however, under the forced division of labour, people in positions of power act out of self-interest, or egoism, implementing rules and engaging in practices that protect their own favoured positions, while constraining other individuals in roles that are unsuitable and unfair, given their abilities and interests. This forced situation still provides a degree of solidarity, but "only an imperfect and troubled solidarity," threatened by strain and eventual collapse (Durkheim, 1893: 376). In such a system, the good of society and of most individual citizens is subordinated to the selfish ends of a few.

Here, too, we can discern significant parallels between Durkheim and Marx. It will be recalled that Marx viewed the division of labour between owners and workers as the basis for an inherent struggle in capitalism, one that occasionally erupts into open conflict but that usually bubbles below the surface of everyday life. Similarly, Durkheim sees the forced division of labour as conducive to a sustained struggle between classes. It is most obvious in "class-wars," wherein the "lower classes" seek to change the role imposed upon them "from custom or by law," to "dispossess" the ruling class, and to redress "the manner in which labor is distributed" (Durkheim, 1893: 374). Typically, however, the struggle is less open or visible, for "the working classes are not really satisfied with the conditions under which they live, but very often accept them only as constrained or forced, since they have not the means to change them" (Durkheim, 1893: 356).

It is apparent from these statements that Durkheim's assessment of the division of labour is neither an apology for nor a defence of the existing economic and political arrangements in capitalist society. As long as the division of labour contains these forced elements, in which the powerful employ "express violence" or indirect "shackle," we have an externally imposed structure that serves neither the individual nor society as a whole (Durkheim, 1893: 377, 380). Such "external inequality" must be eliminated because it "compromises organic solidarity" and therefore threatens the very existence of contemporary society (Durkheim, 1893: 379). Durkheim believes that if this elimination requires open conflict in some cases, then so be it. The worst injustice in that event would be "the making of conflict itself impossible and refusing to admit the right of combat" (Durkheim, 1893: 378). Such statements by Durkheim illustrate, as noted earlier, that he generally held liberal-democratic political views and was not the conservative thinker that some have suggested (see also Lukes, 1973: 271; Giddens, 1986: 23; Parkin, 1992: 59).

The Role of the State

Having just noted Durkheim's acceptance of conflict as one means to produce progressive change, we should also acknowledge that his decided preference is for more peaceful solutions. In Durkheim's view, the most serious problems of anomie and forced inequality can be eliminated, or at least reduced, without such radical actions as full-scale revolution or the utter destruction of capitalism (Durkheim, 1896: 204). On this point, then, Durkheim and Marx have different views.

But what are Durkheim's suggestions for dealing with these abnormal features of modern society? Both anomie and forced inequality are problems over rules, the former a result of insufficient regulation and the latter a consequence of unfair regulation. Hence, Durkheim seeks an initial answer by considering the structure in which most rules and laws are generated and administered: the political system, or state. We discussed earlier the tremendous growth in the number and complexity of rules that accompany the expanding division of labour, and how these rules are increasingly codified as formal laws. The state is in effect the structural embodiment of this modern system of "administrative law" (Durkheim, 1893: 219). Durkheim's discussion here has a strong resemblance to Weber's. Like Weber, Durkheim identifies the state as the "central organ" of modern society, expanding, differentiating, taking on a wide range of duties and a "multitude of functions": administering "justice," "educating the young," and managing such diverse services as "public health," "public aid," "transport and communication," and "the military" (Durkheim, 1893: 221–222; 1896: 43; see Weber, 1922: 655, 905, 908–909).

It is logical that this state apparatus, which creates and implements the formal rules of conduct in all of these spheres, should also be the key structure for ensuring that the rules are moral and just (see Sayer, 1991: 80). The appropriate values of individualism, responsibility, fair play, and mutual obligation can be affirmed through the policies instituted by the state in all these fields. Thus, for example, citizens may be socialized by the education system, the communications media, and so on, to embrace moral precepts and to regulate their behaviour toward others accordingly. Under normal conditions, state leaders will serve as moral examples for the population, underscoring "the spirit of the whole and the sentiment of common solidarity" in their own conduct (Durkheim, 1893: 227, 361–362).

Inevitably, of course, not every individual will adopt or accept the moral code. In those presumably rare instances in which criminal or other elements act against the collective interest, the state has the final recourse of physical coercion, through the police or military, with which to exact compliance (Durkheim, 1893: 222–223).

The State, the Economy, and the Occupational Groups

Interestingly, there is one essential function in society that Durkheim believes cannot be infringed on by the state, and that is the economy. The economy is an exception because it has become so sophisticated and complex under the advanced division of labour that only economic specialists can operate it. These are the people who can deal with the "practical problems" that "arise from a multitude of detail"–problems that "only those very close to the problems know about" (Durkheim, 1893: 360). State leaders can remind economic leaders of their moral obligations to society and can moderate excesses or correct abuses in some cases. However, for the most part, the impact of state leaders on economic life is "vague" and "intermittent" (Durkheim, 1893: 216, 222, 360–361). The specific everyday operation of the economy "escapes their competence and their action" (Durkheim, 1902: 5). Elsewhere Durkheim implies that this situation could change in the future. The growth of economic centralization and big industry, a recent development in Durkheim's time, could lead to conditions that would require state intervention in the economy, if the state apparatus were itself sufficiently developed to take on the task (Durkheim, 1896: 42–43). Otherwise, however, the state is not equipped to become extensively involved in industrial or commercial activities "without paralyzing them" and impeding their "vital" function (Durkheim, 1893: 239).

This exclusion of the state's moral safeguards from the economic sphere poses serious problems for Durkheim and threatens the prospects for a normal division of labour. For how can we alleviate social injustice and social strife if the economic structure is exempt from the moral constraints of the state? How is it possible to eliminate forced inequality and class conflict when the main arena in which such struggles are fought cannot be regulated?

Durkheim's proposed solution to this dilemma is a system of "professional groupings," or "occupational groups" (Durkheim, 1896: 203; 1902: 28–29). The exact nature of these organizations is not certain, but they are clearly not labour unions or occupational associations in the usual sense. That is, their role is not to promote the special interests of particular trades or professions; rather, they are supposed to foster the general interest of society at a level that most citizens can understand and accept (Durkheim, 1902: 10). The modern state has become "too remote" in many cases, "too external and intermittent to penetrate deeply into individual consciences" (Durkheim, 1902: 28). However, the occupational groups are situated between the state and the individual, making them capable of affirming in a more immediate fashion those moral principles that the state can foster in only a general and abstract way.

Their special position means that the occupational groups are also particularly suited to dealing with the problem of moral regulation in the economic sphere. Unlike the state agencies, the occupational groups are able to understand the workings of the economy, because they are a part of it. Thus, they can appreciate the problems of business leaders and individual workers alike. They can therefore play a pivotal intermediate role, impressing on both owners and workers the need to act with moderation and mutual respect in their dealings with each other. The overall effect should be a morally regulated economy, a normal division of labour, and a reduction of class conflict (Durkheim, 1902: 10, 14, 31).

Future Society

It is apparent from the preceding discussion that Durkheim is generally optimistic about the society of the future. His goal of social solidarity based on morality and justice can be accomplished, he believes, provided the capitalist division of labour is adapted to rid itself of anomie, egoism, class conflict, and forced inequality.

In part, Durkheim expects that these abnormalities will be alleviated by the natural evolution of the division of labour. The division of labour, in his view, is already responsible for easing similar problems in traditional feudal and caste societies. Those "prejudices" that once gave favoured rank to people merely because of aristocratic lineage or ascribed religious superiority have gradually been "obliterated" by the division of labour and its need for special, individual talents (Durkheim, 1893: 379).

In much the same way, the "special faculties" required for filling important positions in modern capitalism make it increasingly difficult to obtain these positions solely through inherited class privilege (Durkheim, 1893: 312). To be sure, some inheritance does still occur and unfairly restricts those who are deserving but are born "bereft of fortune" (Durkheim, 1893: 378). Nevertheless, the growing popular sentiment is for the end of such externally imposed inequalities. People have come to believe that inequality should not arise from any outside force but instead should reflect differences in individual "merits"—the talents and efforts that each person contributes in the service of society (Durkheim, 1893: 379, 407–408).

The belief in external equality, or in what might be called equality of opportunity, is so widespread that it "cannot be a pure illusion but must express, in confused fashion, some aspect of reality" (Durkheim, 1893: 379). Of course, the agencies of society should not remain passive while this trend emerges but should actively encourage it. The state, in particular, should institute policies that promote external equality. One key means for doing so is to abolish the inheritance of wealth and property. Economic justice, based on "just contract" between free individuals and a "just distribution of social goods," cannot occur "as long as there are rich and poor at birth" (Durkheim, 1902: 29; see also Durkheim, 1893: 388). In addition, Durkheim suggests that various kinds of social welfare should be provided to "lighten the burden of the workers," to "narrow the distance separating the two classes," and to "decrease the inequality" (Durkheim, 1896: 57).

Such measures, in Durkheim's view, are important if we are to achieve a normal division of labour, one that is free of forced, external inequalities and capable of promoting the solidarity he envisions. Nevertheless, economic justice alone is not enough. In fact, we may even endanger social solidarity if we become too preoccupied with such issues. That is, even if our future society is one in which people "enter into life in a state of perfect economic equality," we have other concerns that are more crucial and that cannot be ignored. These are the more pervasive "problems of the environment"—not just our rights within the economic sphere but also our important "duties toward each other, toward the community, etc." (Durkheim, 1902: 30).

For Durkheim, to harbour an excessive concern with the distribution of economic rewards is to mistake "the secondary for the essential" (Durkheim, 1896: 57). Such a concern puts priority on individual interests rather than on the collective good and each person's obligations to society as a whole. Once again we see the prime significance that Durkheim attaches to moral regulation of all citizens, by all citizens, and for all citizens. If true morality is heeded, the issue of economic justice will tend to be resolved along with society's other problems. If, however, the concern over economic matters takes undue precedence, individualism will deteriorate into egoism and self-interest, human desires for gain or power will override moral power, and the "law of the strongest" will be the only law (Durkheim, 1902: 3; see also Durkheim, 1896: 57, 199–200). Durkheim's perspective on freedom is particularly significant in this regard. Freedom is not the absence of constraint and cannot not exist if moral and legal rules are eliminated from the economy and other spheres of social life: "Quite on the contrary, liberty … is itself the product of regulation. I can be free only to the extent that others are forbidden to profit from their physical, economic, or other superiority to the detriment of my liberty. But only rules can prevent [such] abuses of power" (Durkheim, 1902: 3; see also Peyre, 1960: 21).

Durkheim's projections and proposals concerning the nature of future society reveal some important similarities to, and equally important differences from, the projections made by both Marx and Weber. In the concluding sections on Durkheim, we will consider some of these major points and their significance for the analysis of social inequality.

Durkheim and Marx

There are several aspects of Durkheim's discussion that, on the surface at least, sound intriguingly Marxian. In particular, a comparison of his normal division of labour with Marx's first or transitional stage of socialism reveals some noteworthy parallels: the elimination of property inheritance, though not of all property ownership; the removal of unequal opportunity from the division of labour, though not necessarily of unequal rewards and positions; and the expansion of the state's role as administrator of social services and guardian of the collective good (see Marx and Engels, 1848: 94; Marx, 1875: 8–9).

Of course, these broad similarities cannot mask certain differences between the two men's views. Marx's future society, even the temporary socialist phase, takes us much further from contemporary capitalism than does Durkheim's normal division of labour. Despite the existence in Marx's socialism of a state administration, unequal rewards, and some private property, the impact of such structures and processes is supposed to be curtailed to a much greater extent than in Durkheim's projected society. Moreover, even these vestiges of capitalist society and bourgeois consciousness will disappear, for Marx, when we move beyond transitional socialism and achieve true communism. In contrast, Durkheim clearly is less concerned that capitalism be radically restructured. The current system may take on quasi-socialist elements, but this should occur primarily through gradual modification and adjustment: "It is not a matter of putting a completely new society in the place of the existing one, but of adapting the latter to the new social conditions" (Durkheim, 1896: 204). Besides, in the last analysis, it is not the precise nature of the economic system but rather the problem of social solidarity that, for Durkheim, is the primary concern.

It is interesting that, even on this last point, Marx and Durkheim are again somewhat similar. Both writers desire a society of united individuals, where people interact in a spirit of mutual responsibility, contributing in their own way to the collective interest. Both also believe that problems such as the *distribution* of rewards and resources to individuals can be solved if the social *relations* between people are first geared toward the general well-being of society and all its members. Thus, both theorists really embrace the same primary goal of social solidarity and remain hopeful that it will be achieved. Their disagreement concerns mainly the means or preconditions for its achievement. Whereas Durkheim expects that the collective good can be attained within capitalism, Marx believes it can never be achieved without capitalism's prior demise. Hence, as one analyst has observed, Marxism "finds Durkheimian theory much to its taste *after* the revolution" (Parkin, 1979: 180).

But an even more basic disagreement arises from their assumptions about human nature. Durkheim's hopes for the good society are always tempered by his conviction that humanity, while not inherently evil, does require the prior constraint of moral regulation to keep egoistic desires in check: "Human passions stop only before a moral power they respect" (Durkheim, 1902: 3, 15). Otherwise, in the struggle between egoism and individualism, "the two beings within us," egoism will win out (Durkheim, quoted in Giddens, 1986: 11). For Marx, however, the problem is almost the reverse; that is, structural constraints, at least within capitalist society, are the greatest threats to the general well-being of people. Left to develop freely, human nature is essentially good, social, and oriented to the collectivity. It is the removal of unjust constraint through the overthrow of capitalism that will permit this natural condition of humanity eventually to emerge.

Durkheim, Weber, and the Problem of Power

Any attempt to compare Durkheim and Weber on the subject of social inequality eventually leads to the problem of power. One basic similarity between the two writers is that Durkheim, like Weber, perceives modern society primarily as a pluralist power system, one in which there are several distinct substructures possessing distinct powers and jurisdictions. While this pluralism is not often explicit in Durkheim's work, it may be inferred from at least two points discussed earlier.

First, there is Durkheim's delineation of the several interdependent functions in the modern division of labour. The economic, political, religious, and other functions correspond to the various substructures that compose society, each with its own sphere of influence and duties to discharge.

The second indication of pluralism in Durkheim's analysis is his expectation that certain occupational groups will emerge in future society. These groups will have their own special roles, mediating between the state and individual citizens and arbitrating economic problems, all in the general interest. Moreover, as separate public organizations, they will have powers that are independent both of private economic interests and of state control (Durkheim, 1896: 203–204; 1902: 7–8). Hence, they would add to the diversity of power bases in modern times, providing some protection against abuses of power, especially by those controlling the economy and the state.

If there is one important difference between Durkheim and Weber on the question of power, it concerns the disposition of the state or political structure. Like Weber, Durkheim recognizes the increased role of government in modern life, because of the expansion over time of legal rules and the attendant growth of a state apparatus to create and administer these regulations in a wide range of activities. Nevertheless, a crucial divergence between the two writers is Durkheim's relative unconcern with what is Weber's greatest fear: domination and oppression by the state bureaucracy (see Parkin, 1992: 38–39). Here Durkheim resembles Marx somewhat, for both assume that state leaders, under the proper conditions, will be administrators pure and simple, acting out of the same selfless regard for the collectivity as other citizens and posing no threat to democracy or freedom.

For Weber, however, the good will of state bureaucrats cannot be assumed. Those in a legal position to protect the general interest can, instead, use their position to the detriment of the general interest. Like Weber, Durkheim no doubt hoped that the pluralist nature of modern society would provide a set of countervailing forces to control such power abuses. Perhaps the occupational groups, for example, could serve as monitors against improper actions by the state and other structures (see Giddens, 1986: 28). In a manner similar to that alleged for the free press or for public-interest pressure groups, they could observe wrongdoings, alert the citizenry against them, and mobilize public support to correct them.

In the end, however, Durkheim's discussion of the entire question of state power is incomplete. This is perhaps understandable, given his basic assumption that the struggle for power is not the principal process shaping social life. Such problems as class conflict and the abuse of power clearly exist, but they are really symptoms of more basic ailments caused by insufficient morality and solidarity. Durkheim maintains his conviction that the members of society are united by more than the struggle between interest groups. Interaction based purely on shared interest is inherently unstable: "Today it unites me to you; tomorrow it will make me your enemy" (Durkheim, 1893: 203). Society involves more than a contested activity, with shifting alliances competing for power and rewards. If it is not more than this, society will disintegrate, for then "the state of war is continuous" (Durkheim, 1902: 6).

SUMMARY

In this chapter, our principal focus has been on the thought of Émile Durkheim. We have seen that Durkheim is keenly interested in identifying the sources of solidarity in human societies and have found that, in his view, the division of labour serves as the prime mechanism for providing social solidarity in modern times. Our reading of Durkheim's discussion of the division of labour and related issues also has shown that he has a clear understanding of many important concerns in the field of social inequality, including problems of class conflict, injustice, and the potential abuse of power by privileged groups in society. We noted several interesting similarities, as well as certain differences, between Durkheim's views on such questions and those put forth by both Marx and Weber. In general, we have found that Durkheim's treatment of the crucial issues in social inequality is an optimistic one, as reflected in his expectation that modern societies will continue to improve and progress, provided that we work on removing the various abnormalities in the division of labour as it currently exists. In the next chapter, we shall find that a subsequent school of thought, structural functionalism, not only shares Durkheim's optimism about the prospects for inequality but also argues that the inequalities that do exist in advanced societies can actually serve a positive purpose for the people who live in them.

CHAPTER FIVE

Structural Functionalism and Social Inequality

> *The whole social structure, the whole system of positions, may be viewed as a legitimate power system.*
>
> KINGSLEY DAVIS
> *Human Society,* 1949

INTRODUCTION

In this chapter, our task is to assess the fourth major conception of inequality that has emerged from sociological theory. This conception involves the general orientation known as *structural functionalism.* In contrast to Marxian or Weberian theory, for example, structural functionalism is not clearly identified with any single thinker; hence, its origins are not simply traceable to one source. Nevertheless, it is apparent that structural functionalism grew primarily out of a general trend toward functionalist analysis in the nineteenth century—a trend that found disciples in such diverse fields as biology, art, law, and architecture (Kallen, 1931: 523–525).

Among the social sciences, structural functionalism first achieved prominence in anthropology, especially in the work of Alfred Reginald Radcliffe-Brown (1922, 1935, 1948, 1952) and Bronislaw Malinowski (1926, 1929). Its beginnings in sociology can actually be found somewhat prior to these writers, as evidenced in some of Durkheim's early work near the turn of the century (e.g., Durkheim, 1893, 1895; see Lukes, 1973: 138, 527–528). However, structural functionalism did not enjoy its greatest prominence in sociology until the 1940s and 1950s, with the publication of several influential works by a number of American sociologists, most notably Talcott Parsons (e.g., Parsons, 1940, 1947, 1951, 1953; see also Davis and Moore, 1945; Merton, 1949; Davis, 1949).

Because structural functionalism is a school of thought rather than a single theory, it is difficult both to define it precisely and to review its principles in a manner that everyone will accept. There are many differences in conceptual definitions and theoretical emphases within the rather broad range of writers who make up this school (see Alexander, 1985, 1987; Alexander and Colomy, 1990). As a preliminary statement, however, we can say that structural functionalism is characterized by a particular strategy of inquiry. This strategy entails the investigation of society as if it were a system of parts that are interconnected to form various *structures*, each of which fulfills some *function* for the system.

We will examine the meaning of these two terms in more detail later, but for now we can note that the concept of structure used by structural functionalists is quite similar to that implicit in the formulations of Marx, Weber, and Durkheim: an organized pattern of relationships among individuals or social positions (e.g., Parsons, 1951: 21; Johnson, 1960: 48; Williams, 1960: 20; Levy, 1968: 22). In fact, this definition of structure is similar in general terms to that used by most sociologists today (but see also Lévi-Strauss, 1968; Giddens, 1979; Sewell, 1992). It is the idea of function that really distinguishes structural functionalism from the rest of sociology. Unfortunately, certain disagreements have arisen over the precise meaning of function. In Durkheim's early view, a function is the "need" that a structure fulfills for society (Durkheim, 1893: 49). Some more recent writers define function as a simple "consequence" or "result" of a particular structure's operating in a system (Parsons, 1951: 21; Levy, 1968). This definition leaves unsaid whether or not the function is planned or intentional. Perhaps a more forthright approach is to treat a function as the anticipated or expected "contribution" that a structure makes to society or one of its subsystems (Fallding, 1968: 77–78).

The diverse and sometimes vague meanings attached to the idea of function are a primary source of confusion and dispute between structural functionalists and their critics. In fact, this allegation of vagueness and imprecision has been applied to the whole structural-functionalist perspective. Like other grand theorists, it is argued, structural functionalists attempt to fit all social phenomena into one scheme, thereby making their analysis too abstract to be meaningful. Important details and features of society are inevitably excluded or de-emphasized, especially those elements that are not easily explained by the perspective. Thus, in the case of structural functionalism, critics claim that very little is said about such issues as the roles of class and power in social life, the positive part that conflict can play in human interaction, and the beneficial aspects of social change (see Mills, 1959: 35–42; Lockwood, 1956; Dahrendorf, 1958, 1959; Wrong, 1961).

Some structural functionalists have attempted to show that their framework is capable of including such concerns and of seeing beyond vaguely defined functions or general system needs (e.g., Parsons, 1966; Fallding, 1968). Nevertheless, the conviction remains among many observers that structural functionalism ignores essential features of society or pays only lip-service to them (see Demerath and Peterson, 1967). One of the best-known modern thinkers in the functionalist tradition, while contending that such criticisms are often one-sided, has also acknowledged the deficiencies of functionalist theory, including its implicit aversion to social instability or change and its insufficient concern with the problem of power differences (Alexander, 1985: 14; 1987: 102, 107, 124). Some of Parsons's work, in particular, has been used to illustrate these difficulties in the structural-functionalist perspective (e.g., Alexander, 1990b: 8–10; Camic, 1991; Collins, 1994: 202–203). In spite of such criticisms, however, several recent writers have shown a revival of interest in, and an acknowledged appreciation of, the key insights to be found in the works of Parsons and other structural functionalists (e.g., Mouzelis, 1995: 81; Alexander, 1998: 4–5).

It is not our purpose in this chapter either to examine the debate over structural functionalism in its entirety or to attempt a resolution of it. However, the great influence that structural functionalism has had on social theory makes it essential that we consider the main elements of this perspective and the key criticisms they have raised. Our special concern, of course, is with what structural functionalism has to say about social inequality. If the critics are correct, the exclusion of such issues as class, power, conflict, and change would make the structural-functionalist view of social inequality and its causes quite different from the other theories we have already reviewed. At the same time, however, wherever such criticisms misrepresent the structural-functionalist approach, it may be possible to identify similar, or at least compatible, elements in all of these perspectives.

We begin our assessment of structural functionalism with some observations on the links between Durkheim and the modern structural-functionalist school. It will be argued that Durkheim does provide a basis for structural-functionalist sociology in certain general respects but that his conceptions of specific issues are often significantly different from the ideas espoused by recent structural functionalists. In the remainder of the chapter, the focus will shift to modern structural functionalism, especially the predominant version developed by Parsons (1937, 1951). This discussion will begin with a review of the basic conceptual scheme of structural functionalism. We will then assess the key criticisms of this approach, especially those concerning the concept of function. This discussion of the general orientation will set the stage for our central task: an examination of the famous structural-functionalist analysis of social inequality, especially as developed by Parsons (1940, 1953) and by Kingsley Davis and Wilbert E. Moore (1945). The important points of similarity and difference between modern structural functionalism and the theories of Marx, Weber, and Durkheim will be of special concern throughout this section of the chapter.

DURKHEIM AND MODERN STRUCTURAL FUNCTIONALISM

Structural functionalists claim several theorists among their ancestry, including Weber and occasionally even Marx; however, there is little doubt that Durkheim has had a more significant formative influence on modern structural functionalism than either of these two writers (Parsons, 1937; Fallding, 1968: 54; 1972: 94; Collins, 1985: vii–viii). In particular, the tone and emphasis in current structural-functionalist analyses suggest an image of society that generally resembles Durkheim's. Thus, for example, structural functionalists share Durkheim's conception of society as a systematic aggregation of interrelated parts, all fulfilling important tasks for the common good. Structural functionalists likewise retain Durkheim's special interest in organizing and coordinating these tasks in order to maximize social integration and the chances for societal survival.

In addition, structural functionalists point to some of the same forces that Durkheim stresses when discussing the integration of modern society. First, integration is aided by the functional interdependence of individual social actors engaged in their own specialized tasks. This phenomenon is similar to Durkheim's idea of achieving social solidarity through the division of labour. Second, functionalists see integration stemming from the collective acceptance by individuals of a system of specific norms or rules that guide social relationships and regulate interactions. These norms correspond to the complex set of moral and juridical rules that Durkheim believes will promote the smooth, cohesive operation of modern societies. Finally, structural functionalists argue that integration is enhanced by popular adherence to a body of common values and beliefs. Like Durkheim's collective consciousness, these values are said to pervade social life, providing at least general principles of human conduct and another means for binding society together.

Broad parallels of this sort make it apparent that the similarities between Durkheim's original work and modern structural-functionalist analysis are both noteworthy and genuine. At the same time, however, several important differences are also evident and must be kept in mind (Alexander, 1985: 9; Collins, 1988a; 1994: 202–203, 234–236). While no single factor or person is responsible for these differences, the principal cause is probably to be found in the work of Talcott Parsons.

Parsons played a pivotal role in bringing the writings of Durkheim to the attention of English-speaking, especially North American, sociologists. In his early work, *The Structure of Social Action*, Parsons attempts a reinterpretation of Durkheim, which, in combination with his views of Weber and others, he uses as a basis for his own theoretical perspective (Parsons, 1937). Much of what is now the core of structural-functionalist sociology is really Parsons's attempt to filter Durkheim, and to some extent Weber, through his own conceptual framework.

In order to grasp the essentials of contemporary structural function-alism, and in order to discern its major departures from Durkheim and its other progenitors, it is important that we briefly consider the basic conceptual scheme employed by Parsons and most subsequent structural functionalists. We will then assess how this scheme gives rise to the structural-functionalist analysis of social inequality. In the process, we will discuss the key ways in which the structural-functionalist perspective diverges from the outlooks of Durkheim, Weber, and Marx.

STRUCTURAL FUNCTIONALISM: THE BASIC CONCEPTS

Structure

As the name suggests, structural-functionalist theory portrays society as a system of social *structures*. Structures in this sense are really patterns of relationship or interaction among the various components of society— patterns that are relatively enduring because interactions occur in a regular and more or less organized way.

The structural components of society exist at several levels of generality. At the most general level is society as a whole, which may be viewed as a single, overarching structure. The second level down is a series of more specialized structures that interconnect to form society, rather like the pillars of a building or, following Durkheim, like the organs of a living organism.

Each of these second-level structures is itself characterized by further task specialization. Thus, for example, we can think of the economy as one of these second-level structures. It performs a particular task that is itself a combination of interrelated, even more specialized, tasks: extracting raw materials, such as wood or iron ore; processing these materials into goods, such as lumber or steel; using these goods to manufacture finished products, such as furniture or cars; distributing, selling, and servicing these finished products; and so on. Special economic tasks thus give rise to their own special substructures, and these, taken together, compose the overall economic structure (see Parsons, 1953: 400).

Status and Role

Ultimately, the decomposition of structures in this way leads to the most basic level of analysis: the individual social actor. In the structural-functionalist perspective, each individual occupies a *status* within the various structures of society. Status here does not refer to the prestige of the individual's position, but simply to the position itself. The individual occupying a status is also afforded certain rights and duties, which are the individual's role in that status (Williams, 1960: 35–36). Thus, status and role tend to go together in what Parsons calls the "status-role bundle" (Parsons, 1951: 25; 1953: 393–394).

Social structure, then, is the interconnection of statuses that results when actors perform their assigned roles in interaction with one another. Thus, when those who fill the various worker, owner, manager, and related statuses of society perform their roles, we have an economic or occupational structure. When those who are voters, legislators, government officials, and so on, fulfill their roles, we have a political structure. The same perspective can be employed to characterize the educational, religious, and other social structures that society comprises. One of the unifying aspects of this image of society is that each individual may have a status and a role in all of these structures at the same time. In effect, the individual actor is plugged into numerous structures, rather like a multiple electrical outlet. A related image is to see the individual as segmented into several roles, almost like the slices of an orange (Williams, 1960: 517).

Norms, Values, and Institutions

Under the label *social structure*, some structural functionalists include not only the status-role interactions but also the specific rules and general beliefs, the *norms* and *values*, that regulate these interactions (e.g., Johnson, 1960: 51). A more prevalent view among structural functionalists, however, is that norms and values are not structural but *cultural*, existing in a different conceptual space that overlays social structures. In other words, norms and values are really ideas or symbols that individuals keep in mind as codes and sanctions for their interactions in their various status-roles (Parsons, 1951: 327; Williams, 1960: 20–30).

This disagreement over whether norms and values are structural or cultural parallels the confusion over the meaning of superstructure in Marxian theory, an issue discussed in Chapter 2. Just as superstructure can refer both to sets of ideas and to the structures that embody them, so too are norms and values sometimes equated with the structures that represent them. This equation occurs especially when structural functionalists speak of *institutions*, which are the most permanent, pervasive, and obligatory systems of norms and values in society (Williams, 1960: 30–31; see also Parsons, 1937: 407; 1951: 39; Levy, 1968: 27). There is, for example, a common tendency to equate religious institutions, the system of religious rules and beliefs, with the religious structure, the administrative apparatus that has developed for the nominal purpose of implementing these religious beliefs. The same may be said concerning economic institutions and the economic structure, education and the educational system, or political institutions and the structure of the state (Williams, 1960: 517).

This rather loose equation of ideas and realities, of institutions and social structures, is sometimes a useful device, for it aids in organizing people's thinking about social arrangements. However, it is important to look for cases where stated values or norms and *actual* structural relations do not correspond. For example, a dominant value, such as equal opportunity within capitalism, may not be served by, and may even be impeded by, existing structural arrangements, such as the formal entrenchment of the right to inherit property or wealth.

Function

This brings us to the last key concept in structural-functionalist analysis: the idea of *function* itself. We have noted that there is a correspondence between social structures and the institutions that guide their activity. In a similar way, there is a rough correspondence between these two conceptions and the various functions of society. Thus, for example, a structural functionalist would characterize the economy in capitalism as a structure, or system of structures, operating according to a set of corresponding economic values, norms, or institutions, such as private-property ownership, in the performance of its main function: the provision of the material means of existence for society's members.

But what, precisely, is meant here by the term function? Is this the economy's intended purpose in society, as planned by some agent or leader, or is it simply an unintended consequence of economic activity? Is the provision of the material means of existence the actual contribution the economy makes to society, or is it merely the stated goal attributed to it by some people and not others? Such questions inevitably arise when the term function is used in sociology. Moreover, as noted at the start of this chapter, structural functionalists and their critics alike cannot agree on how to answer them, because of the confusion and disagreement over the definition of function. Problems are further compounded by the use of several ideas that are related to function but are somehow distinct from it, including "functional requisite," "functional prerequisite," "functional imperative," and "functional problem" (Merton, 1949; Aberle et al., 1950; Parsons, 1951, 1953; Johnson, 1960).

Even if it were possible, it is well beyond our purposes here to sort out the conceptual wrangles involved in the idea of function. And, in any case, most of these disagreements do not undermine at least a basic consistency in the use, if not the definition, of the term. For most structural functionalists, a function is really a *social task*, an activity that must be performed with some degree of adequacy if social groupings are to exist and to sustain their members. Among these tasks are a range of operations: socialization and education of the young, administration of economic and political affairs, regulation of criminal behaviour, and so on.

Presumably, most sociologists would agree that such activities are indeed essential to any society (Giddens, 1979: 113–114). Hence, if a straightforward definition along these lines were to be employed consistently, there might be fewer problems with the meaning of the term function. Instead, however, the ambiguity surrounding this central concept in the structural-functionalist scheme has provoked serious suspicions regarding the intentions of structural-functionalist analysis. Among the numerous allegations that have been made, two in particular will be noted here.

First of all, some observers believe that when structural functionalists describe the tasks of society as functions, they are really promoting the view that the existing structures and institutions of society are good or ideal, functioning properly and satisfying society's needs. The implication is that any alteration in the established arrangements must, in their terms, be dysfunctional—that is, disruptive of the stable operation of society. Thus, detractors believe that structural functionalists implicitly adopt an uncritical acceptance of the current social structure, sometimes combined with an outright distrust of social change.

A second key problem involving the idea of function concerns how we decide if something is truly functional or not. Critics argue that structural functionalists judge structures or institutions solely on the basis of whether they meet the needs of society *as a whole*. To critics, judgment on this basis alone implies that a structure or system of rules will be deemed functional as long as it deals with some important societal task, regardless of its consequences for particular groups or individuals *within* society.

For example, an important task for society is to control crime, violence, and other forms of antisocial behaviour. Elements of the political structure, particularly the police and the military, may be in place to perform this task. However, if the people who command these structures employ policies, official or otherwise, that control crime and violence by means of strict curfews, imprisonment without trial or evidence, and so on, the social-control function may be achieved but at tremendous cost to other segments of society. These organizations and rules could become tools not only to regulate crime but also to eliminate legal and peaceful dissent—to oppress political dissidents, minority groups, or other legitimate factions who may not hold favour with the society's leadership. Thus, if we were to judge the functionality of a structure *only* in terms of the vague abstraction called society, the dysfunctional aspects of that structure might be ignored, unintentionally or otherwise. Moreover, it would be possible in such a circumstance to hide what are really the *special* goals and interests of certain groups, particularly those in control of the economy or the state, by representing them as society's goals or the general interest.

One important feature of these criticisms is that they result more from what structural functionalists fail to discuss than from what they actually say, from what they imply more than from what they explicitly advocate. This is not to argue, of course, that all the disagreements between structural functionalism and its critics are simply misunderstandings. As we shall see, there are several fundamental differences, especially in the analysis of social inequality, that seem irreconcilable. However, it is contended here that the concept of function in sociology has become a loaded term, detested by some and embraced by others, primarily because of the implied meanings attached to it over the years. There is nothing inherent in the concept that should provoke such feeling, especially if the term is used mainly as a synonym for social task, as discussed earlier.

It is interesting to note in this regard that some Marxist scholars have been among the strongest opponents of the concept of function as employed by the structural-functionalist school. However, there seems to be nothing intrinsic in the term that explains this opposition. In fact, both classical and contemporary Marxists have been known to use the concept of function (or *Funktion*) themselves (e.g., Engels, 1890a: 490; Marx, 1894: 379; Poulantzas, 1975; Weiss, 1976; Carchedi, 1977; Wright, 1978).

We should also note that at least some structural functionalists have attempted to allay the problems raised by their critics. That is, although structural functionalists admittedly place greater stress on the importance of stability in society, some do try to incorporate an analysis of progressive social change into their discussion. In addition, some structural functionalists do not always judge the functionality of social phenomena in terms of the vaguely defined needs of society as a whole. These writers recognize that one must be explicit about who is being served or not served by particular structures and institutions in a social system (e.g., Johnson, 1960: 70; Levy, 1968: 25; Fallding, 1968: 77–78; see also Alexander, 1985: 14; 1990b).

Only One Sociology?

As the preceding discussion reveals, the structural-functionalist conceptual scheme portrays society as a complex system of social structures that operate under the guidance of institutionalized norms and values in the performance of special, vital functions for society and its members.

Although the terminologies sometimes differ, it should be apparent from the earlier analyses that Marx, Weber, and Durkheim share an interest in these same concerns: how social structures are arranged, how they are regulated or governed, and what they do for (or to) the people who live within them. In fact, it could be argued that virtually all sociologists, past and present, are involved primarily in examining these phenomena.

This very general resemblance between structural-functionalist concerns and those of other sociologists seems to be responsible for a bold claim on the part of some structural-functionalist writers: that there is only one sociology and that it is functionalist. In other words, structural functionalism is not a special version of sociology at all but rather an equivalent term for sociology itself (Davis, 1959; Fallding, 1968: 54–55; 1972: 93; Levy, 1968: 22).

Some aspects of this assertion, on the surface, may seem plausible. To begin with, it is true that the basic sociological approach until recent years has been broadly similar to the structural-functionalist strategy. Sociologists do tend to look at society as if it were a system of interconnected parts, of individuals and groups organized in more or less regular interrelationships. Second, this apparent affinity is compounded by the fact that, as Giddens notes, even structural functionalism's critics often seem to voice their opposition to it using a functionalist vocabulary (Giddens, 1979: 60). A third point that seems to favour the structural-functionalist claim is that a generally similar strategy is used outside the social sciences (Levy, 1968: 22). When other scientists wish to understand how something works, they typically perceive the phenomenon in question as if it were a system of interrelated segments. The component pieces can then be separated, analyzed, and mentally reconstructed in an effort to comprehend both the parts and their interconnections. Medical science, for instance, studies the human body as a complex of cells, organs, and subsystems operating in concert to create and sustain life. Biological science examines plants and animals using similar assumptions. The sciences of physics and chemistry likewise proceed from images of chemical and physical matter as composites of elements and particles organized into a systematic whole.

However, a closer look reveals some basic flaws in the argument that structural functionalism is the universal approach in sociology or in science generally. This assertion seems, in particular, to mistake the lesser for the greater. That is, the vaguely similar strategy used by structural functionalists and other sociologists is taken to mean that all sociologists are structural functionalists, whereas it simply demonstrates that all structural functionalists are sociologists. This reversal introduces some unfortunate misrepresentations. Among structural functionalists, it can produce an unfounded self-assurance, a strong conviction that their particular view is the only one to take (for discussion, see Alexander,

1990a; Camic, 1991). As an aside, we might note that a similar conviction, that their own perspective is the only correct one, sometimes occurs in other theoretical camps as well.

An equally serious misrepresentation in this regard is one that occurs among certain critics of structural functionalism. Some critics automatically assume that any perspective that represents society as a structured system, even in very general terms, is merely another brand of structural functionalism and is therefore to be completely dismissed. The point to stress here is that this assumption is just as mistaken as the assumption that structural-functionalism is or should be the universal approach in sociology. Both views fail to recognize that using a similar strategy of inquiry—one that amounts to sociology itself—does not amount to having the same theory of society. Using this very broad sociological strategy, it is possible to develop numerous theories for how social structures are generated, sustained, and transformed, each of which has a distinct emphasis and draws distinct conclusions. Our task is to determine which of these sociological theories is most useful, or seems most accurate, for explaining the phenomenon under investigation. There is no topic in sociology in which these distinctions between theories are more apparent or more crucial than in the analysis of social inequality. In the remainder of this chapter, we will examine the structural-functionalist approach to social inequality and assess the key ways in which this unique view can be compared with and contrasted to the theories of Marx, Weber, and Durkheim.

STRUCTURAL FUNCTIONALISM AND SOCIAL INEQUALITY

Because there are several variants of the general structural-functionalist perspective in sociology, it is not surprising that there are also several different ways in which structural functionalists have approached the subject of social inequality. Among these formulations, the principal versions are Parsons's early "analytical approach," which he later revised, and the better-known though less comprehensive treatment by Davis and Moore (Parsons, 1940, 1953; Davis and Moore, 1945).

There is no doubt that these analyses diverge on certain points (Münch, 1982: 815–816). Nevertheless, the essentials of the various structural-functionalist approaches to the topic of inequality are generally similar. Thus, we will not attempt a detailed review of what, for our purposes, are minor distinctions among these formulations. Instead, we will proceed with an outline of the major themes held in common by the several structural-functionalist discussions of social inequality. In addition to noting these central themes, our second task will be to assess the key areas of similarity and dissimilarity between the modern structural-functionalist perspective on inequality and that posed by each of the three early writers we have considered: Marx, Weber, and Durkheim. Obviously, the number of possible comparisons is very large; therefore, in the interests of clarity and simplicity, only the crucial points of convergence and divergence will be highlighted. The discussion throughout is organized around three principal issues: the differing conceptions of inequality as class structure or as individual stratification ranking, the relative importance of consensus and conflict in generating and sustaining social inequality, and the parts played by power and authority in social hierarchies.

Social Inequality: Class Structure or Stratification?

The predominant strategy for studying social inequality over the years has been to focus attention on issues of class. However, as an alternative to the class perspective, researchers will sometimes use a different view, one in which society is perceived as a hierarchy composed of layers, or *strata*. In this stratification perspective, individuals are ranked along a continuum or ladder and divided into discrete categories, which are, in effect, the strata for that particular analysis.

The stratification criterion varies depending on the researcher, but in most cases it is some objective indicator of economic rank, such as income, education, or occupational level. Often researchers will attempt to combine these separate rankings into some overall hierarchy. Perhaps the most common procedure is to calculate a single score for every occupational title, based on the average income and education of those engaged in each occupation. Thus, the occupational structure is transformed into a scale of overall "socioeconomic" rank (e.g., Blishen, 1967; Blau and Duncan, 1967; Blishen and McRoberts, 1976; Pineo, Porter, and McRoberts, 1977; Blishen, Carroll, and Moore, 1987). Another approach is to rely on the *subjective* assessments, by representative samples of the population, of the general *social standing,* or *prestige,* of occupations. These occupational-prestige scores have also been used as indicators of the individual's overall stratification position (e.g., North and Hatt, 1947; Inkeles and Rossi, 1956; Hodge, Siegel, and Rossi, 1964; Hodge, Treiman, and Rossi, 1966; Goldthorpe and Hope, 1974; Treiman, 1977).

These various forms of the stratification perspective can be quite useful for students of social inequality. To take a simple illustration, imagine that we wished to determine the degree of income inequality in society. We could proceed by ranking people according to their annual incomes, dividing them into deciles (ten strata of equal size), and then comparing the total amounts of income earned within each stratum. If we determined, for example, that the top tenth of the population receives 50 percent or more of the total and the bottom tenth earns only 1 or 2 percent, this would suggest an extremely unequal society in terms of the distribution of income.

Stratification research along these lines has been done extensively in sociology. In fact, some researchers use this approach almost exclusively; however, it is also common for the same researcher to employ a class view at certain times and a stratification view at other times, depending on the researcher's purposes or interests. For example, some Marxist scholars, despite being concerned primarily with class issues, will occasionally take a stratification perspective in their investigations (e.g., Kolko, 1962; Johnson, 1979; see also Wright, 1994: 89–91).

The structural-functionalist approach to social inequality is really a special version of this stratification perspective. Structural functionalists also conceive of inequality as a general, continuous hierarchy along which individuals can be ranked (Barber, 1957: 77). In their conception, people are stratified on the basis of the various status-roles they perform in society. People tend to fulfill numerous status-roles in life, so that one's ranking is "the general resultant of many particular bases of evaluation" (Parsons, 1951: 132). In most cases, however, structural functionalists focus on the status-role the individual holds in the occupational sphere. Occupation is seen as the best single indicator of general stratification rank, partly because it correlates with many other bases of ranking, such as income and education, but mainly because it is, for most people, the "functionally significant social role" one plays in society (Barber, 1957: 171, 184–185).

What this approach means is that, for structural functionalists, the stratification system is really a consequence of collective *judgments* by which society (presumably, people in general) *evaluates* the worthiness of a person's occupational position with regard to its importance or contribution to the collectivity (Parsons, 1940: 76–77; 1953: 386–387). This approach corresponds loosely with the occupational-prestige variant of the stratification perspective noted earlier, for both views stratify society in terms of a *subjective* evaluation by others of one's prime social role or status. Of course, it should be emphasized that all prestige researchers are by no means structural functionalists. Nevertheless, this idea of prestige ranking does appear to be the crucial element in structural-functionalist conceptions of stratification (e.g., Davis and Moore, 1945: 242; Parsons, 1951: 132; Barber, 1957: 73; Johnson, 1960: 469–470; Williams, 1960: 97).

That structural functionalists perceive inequality in terms of the value to society of one's occupational position is, of course, consistent with their general viewpoint, especially their overriding concern with collective sentiments and societal needs. This concern is not in itself particularly objectionable. However, as we shall discuss later in this chapter, one may wonder whether it is possible to achieve collective agreement on which occupations are more important or worthy than others. Furthermore, if collective agreement is not possible, who makes these crucial ranking decisions?

One of the most serious weaknesses in the structural-functionalist conception of inequality as prestige stratification is that it becomes interwoven and confused with questions of class. These prestige strata—ranked categories of people with similar occupational prestige—are simply equated with the class structure. Class is thus defined as "a more or less endogamous stratum consisting of families of about equal prestige" (Johnson, 1960: 469; see also Parsons, 1951: 172; Barber, 1957: 73). The prestige—and hence the "class status"—of all family members is judged by the occupation of the household head, who is usually the "husband-father" (Parsons, 1953: 426–427).

Conceptual problems arise when the stratification and class perspectives are conflated in this way, when strata are erroneously equated with classes (Stolzman and Gamberg, 1974). While the two perspectives focus on similar topics, especially economic inequality, they differ in certain crucial respects. For one thing, a stratum, unlike the original idea of class, is not meant to be studied as if it were a real *group,* a set of people interacting with one another or having some sense of common affiliation. Of course, classes are not always real groups either, but there is at least the possibility that classes will form groups in certain circumstances. For example, a precondition for the overthrow of capitalism in Marxian theory is the development of a revolutionary working class—individuals who, in addition to their common economic position, have a common consciousness, a sense of group solidarity, and a collective will to mobilize for political and social change. In contrast, when stratification analysts examine socioeconomic ranks, income deciles, or other types of strata, these cannot be, and are not intended to be, real groups. Instead, they are statistical aggregates, categories of individuals lumped together for particular research purposes.

Second, class and stratification analysts tend to emphasize different aspects of the inequality they study. On the one hand, class analysts are interested primarily in the *relational* consequences of inequality—that is, in the domination or exploitation of one class by another and the impact of these relations on social structure or social change. Recall, for example, that Marx viewed the relationship between owners and workers in capitalism as the driving force behind social change because of the conflict inherent in this relationship between classes. On the other hand, stratification analysts typically focus on *distributive* inequalities, on the differential allocation of income, prestige, and other rewards or advantages to individuals in society (see Goldthorpe, 1972; Hunter, 1986; Wright, 1994).

We shall have more to say on these issues later. For now, it is important to note only that the class and stratification perspectives differ in important ways, and that the structural-functionalist approach tends to blur or ignore these differences. The failure to make such distinctions is rather curious. The structural-functionalist view departs completely from Marx's classical treatment of class as defined by relationship to the means of production. Moreover, despite claims by structural functionalists of an affinity between their work and the formulations of both Weber and Durkheim, there is really very little common ground in either case. In one structural-functionalist analysis, it is explicitly argued that Weber adopts a prestige-stratum definition of class (Barber, 1957: 73). However, as should be clear from Chapter 3, for Weber class is primarily an *economic* concept, related to market position and control over goods and services. In Weber's scheme, the concept that comes closest to the structural-functionalist idea of class as prestige stratum is probably the status group. But the prestige strata in the structural-functionalist scheme are not groups, whereas Weber's status groups obviously are. Besides, Weber goes to great lengths to show that status groups and classes are not the same in any case, asserting the conceptual independence of economic class and status honour.

We might expect the structural-functionalist affinity with Durkheim to be very close on this issue of class and stratum, given their broad similarities in general approach, and given the conventional assumption that Durkheim's work is a prototype of modern structural functionalism. However, there are significant disparities. Presumably, Durkheim would agree with the structural-functionalist argument that, ideally, inequality arises through the differential evaluation of what individuals do in and for society. However, this individual stratification has little or no theoretical connection with the idea of class. In fact, in those relatively rare instances where Durkheim speaks of class, his remarks, as we have already seen, are rather close to Marx's and Weber's. His emphasis in discussions of class is, like theirs, on the *economic* differentiation of people, not on their differential prestige or moral worth (see also Grusky, 2005).

Our intention in this section has been to outline the major differences between the class and stratification perspectives on inequality, especially as these relate to certain conceptual difficulties in the structural-functionalist approach. Structural functionalism's particular view of social inequality seems to combine selected and modified elements of traditional class-based theories, which have been grafted on to the more recent stratification school of social research. This attempted amalgamation has really only confused these two important, but quite separate, formulations. The failure to recognize these distinctions is a serious flaw and underscores our earlier conclusion that structural functionalism, despite its claim to the contrary, is not capable of subsuming all of sociology in any satisfactory and complete fashion.

Conflict, Consensus, and Social Inequality

A second important issue to address in any discussion of structural functionalism is the debate over whether society is characterized primarily by underlying conflict and struggle or by general consensus and agreement among its members. More specifically, we need to examine whether it is mainly through conflict or through consensus that individual and group inequalities become established in social structures.

To understand the structural-functionalist position on this issue, and to compare it with the other theories we have examined, a useful strategy is to locate each approach along a continuum according to the relative importance that each attaches to conflict and consensus. On such a scale, Marx probably would fall closest to the conflict end, with Weber nearby. Durkheim would be some distance away from Marx and Weber, but not at the consensus extreme. The modern structural-functionalist school would lie closest to the consensus pole. Let us briefly examine the reasoning behind each of these placements.

Marx, Weber, and Durkheim

Marx's conflict emphasis is most evident in his central premise that unequal relations of power and privilege are the products of a continuous historical struggle for control of the means of production. Although there may be extended periods of time when open social unrest does not occur, this relative stability is not a sign that there is a general consensus about the justice of existing social inequalities. In most cases, this apparent acceptance is a mere illusion and instead reflects a variety of other processes, including successful ideological manipulation of the population by the ruling class, lack of awareness in the lower classes about the causes of and remedies for their subordinate position, or simple despair in the lower classes that inequities can ever be alleviated.

Compared with Marx, Weber seems somewhat more likely to conclude that there is some general agreement about the justice of social inequality. To a limited extent, at least, individuals compete with each other on the basis of generally accepted ideas about the exchange of goods and services in the economic marketplace. Most people also concede that bureaucratic hierarchies and other institutionalized inequalities are a fact of modern life. Nevertheless, claims of consensus should not be overstated on these grounds. First of all, to say that one accepts the inevitability of inequality is not to say that one agrees with how particular hierarchies arise. Moreover, the ultimate origins of this acceptance should not be overlooked. For Weber, from the earliest times the monopoly of physical force has been a crucial factor in generating inequality and the domination of one group by another. This coercive aspect of inequality may become less obvious as societies develop and power is formally institutionalized, but the ability to force compliance from subordinates nevertheless remains a key underlying basis of social inequality.

Thus, to Weber, considerable conflict and antagonism often lie at the root of what on the surface are stable social hierarchies. Some citizens will actively embrace these existing social arrangements, but others will abide them for a mixture of quite different motives: habit, custom, fear, or a failure to discern alternatives. Such motives have little to do with a general consensus about the justice of the social order. Moreover, even in those societies where there is evidence of harmony and agreement, social action regularly includes a struggle for advantage among opposing factions and competing interest groups.

Analysts tend to identify Durkheim more with a consensus view of society, and less with a conflict perspective, than either Marx or Weber. This view probably stems from Durkheim's concept of the collective consciousness, the commonly held values and beliefs that he perceives as a unifying force among early peoples. While there is little doubt that Durkheim is interested in such consensual aspects of society, we should not overlook the departures from consensus that he also perceives in social structures. First, we should remember that Durkheim sees the collective consciousness as necessarily weaker in modern times. To be sure, shared sentiments, such as a belief in freedom or equal opportunity, may still set a general moral tone for social action. Beyond this, however, consensus is unlikely, because the growing division of labour makes the legal and moral guides for conduct too specialized, elaborate, and complex for everyone to agree on or even comprehend.

Thus, although Durkheim envisions a just and moral future society, it will not be characterized by some thoroughgoing consensus. What accord there is will stem mainly from *differences,* stemming from interdependence among people in the division of labour, and from the recognition by most citizens of the need for mutual cooperation and obligation. We should also note that for Durkheim this differentiated basis for social harmony, the normal division of labour, "is far from being on the verge of realization" (Durkheim, 1893: 408). In the interim, forced inequalities and class antagonisms persist beneath the orderly façade of present-day societies, calling into question the view that social hierarchies somehow arise out of a collective consensus among the people.

Structural Functionalism

It is here that structural functionalism enters the discussion. It would be inaccurate to claim that the entire structural-functionalist school completely ignores the existence of conflict in its portrayal of society (see, e.g., Davis, 1949; Fallding, 1968, 1972; Alexander, 1985, 1987). Nevertheless, there is little doubt that structural functionalism deals with conflict mainly as a secondary issue and that, more than any other perspective, it sees consensus as the principal foundation of social structures.

On this basis, for example, Parsons disagrees with Durkheim's view that moral consensus, as reflected in collectively held beliefs, has waned in modern times. Instead, Durkheim's collective consciousness has simply changed character and is now embodied in the "ultimate value system" (Parsons, 1937: 400–401). This value system is "inculcated from early childhood" into the individual personality and fundamentally shapes even those specific normative rules, rights, and duties that govern social action (Parsons, 1940: 73–74).

Hence, as social structures develop, their inherent properties tend, on the whole, to be consistent with a collectively held value system. According to structural functionalism, one such inherent property of society is the existence of inequality or stratification. Stratification is a universal aspect of social life, something that has occurred in virtually all known societies (e.g., Davis and Moore, 1945: 242; Davis, 1949: 366; Parsons, 1951: 188; Williams, 1960: 88). To structural functionalists, this prevalence of inequality is evidence of its inevitability and of its acceptability, at least in principle, to social actors.

Even more important to the consensus argument is the claim that there is a high level of agreement in the population about how specific status-roles or occupations should be ranked. At times, structural functionalists concede that people's judgments will not correspond perfectly (e.g., Parsons, 1953: 390). Nevertheless, it is argued, consensus on stratification rankings is quite high, at least in stable, democratic societies (e.g., Parsons, 1940: 71; 1953: 388; Davis and Moore, 1945: 242; Williams, 1960: 93).

And what of the criteria for ranking occupations? What are they, and to what extent is there general agreement about them? As discussed previously, structural functionalists envision stratification as a subjective scale of prestige, social standing, or moral evaluation. Various other factors can influence this evaluation of occupations or status-roles, including the power, possessions, or family backgrounds of incumbents (Parsons, 1940: 75–76; 1953: 389–390; Davis and Moore, 1945: 244–248; Barber, 1957: 30–48). For the most part, though, one's worth to society is the key factor, and it is estimated by considering two overriding concerns: the "functional importance" for society of one's occupation and the "differential scarcity" of people with the talent or training needed for its performance (Davis and Moore, 1945: 243–244; see also Davis, 1949: 368; Parsons, 1953: 403, 410). In other words, jobs such as doctor or scientist, which are allegedly both more important and more difficult to fill than jobs such as dishwasher or waiter, receive higher rankings in the stratification system. Here, too, people are said to be in considerable agreement in their judgments, even for the numerous and complex set of occupations in the modern division of labour (but see Davis and Moore, 1945: 244).

One might ask at this point whether there is any evidence to support these allegations of consensus in the stratification rankings that people make. The most commonly cited evidence is the strong correlation (frequently above .90) that various researchers have reported among occupational-prestige rankings done by samples of respondents in different countries or at different times in the same country (e.g., Inkeles and Rossi, 1956; Barber, 1957: 105–106; Lipset and Bendix, 1963: 14; Hodge, Treiman, and Rossi, 1966; Treiman, 1977).

However, we should note that these results have been criticized on methodological and statistical grounds. For one thing, researchers derive prestige scales by combining the disparate ratings of a sample of respondents, thereby averaging out significant disagreements among respondents in their ratings of specific occupations (e.g., Nosanchuk, 1972; Stehr, 1974; Coxon and Jones, 1978, Guppy, 1981, 1982; Zhou, 2005; for debate, see Balkwell, Bates, and Garbin, 1982; Hodge, Kraus, and Schild, 1982). Thus, it appears that the high correlations among such scales really mask considerable disagreement in the popular evaluation of occupational status-roles. Therefore, while it is unlikely that the observed correlations are totally the result of such measurement problems, it is also unlikely that, beyond certain obvious distinctions, there exists a general stratification scale that most people will agree on (Parsons, 1940: 86; see also Kelley and Evans: 1993).

Before concluding our discussion of conflict and consensus, we should consider one additional issue. Let us imagine, for the sake of argument, that there is a consensus on the way that occupations rank with respect to prestige, moral evaluation, functional importance, or scarcity. We might still wonder why this scale should also give rise to other inequalities—for example, in access to such material advantages as income, wealth, or property. That is, is it not possible that the rewards of prestige and recognition for service to the collectivity might be a sufficient distinction for the most deserving positions, thus eliminating the need for material and other inequalities (Tumin, 1953)? The response to this question varies somewhat, depending on which structural functionalist one consults. On the whole, however, the structural-functionalist view is that material inequalities will occur and that, in stable societies at least, these differences *should* correspond generally to the scale of evaluation.

This view appears to stem from a particular conception of human nature and motivation (Wrong, 1959: 774; Wesolowski, 1966). The structural-functionalist assumption here is that most social actors are oriented to the good of the collectivity, but also to their own "self-interested elements of motivation" (Parsons, 1940: 73; Davis, 1953). Stratification serves the function of satisfying both these needs, one collective and one individual, at the same time. However, satisfying both needs can occur only if the material and evaluative hierarchies tend to correspond. According to

Davis and Moore (1945: 244), unequal material rewards motivate the best qualified and most talented to take on those positions that are the toughest, most important, and hardest to fill. Any distribution of economic rewards not based on an occupation's contribution and worth to society will only act as a disincentive to an efficient division of labour, especially because it will discourage rare and able people from assuming the sacrifices and responsibilities of high office (see Parsons, 1953: 404–405). Note again the implicit assumption of consensus here, this time about the perceived justice of unequal rewards for service and the injustice of alternatives. Especially in Western societies like the United States or Canada, these inequalities are seen as consistent with popularly held beliefs in such values as efficiency, freedom, achievement, and equal opportunity (e.g., Parsons, 1953: 395–396; Williams, 1960: 415–470).

In the end, however, structural functionalists believe that people see these material incentives and economic rewards as of secondary priority anyway, at least in comparison with the primary rewards of prestige and recognition from others. Wealth, while important in its own right, really has a *symbolic* meaning as an index of achievement and high evaluation in society (Parsons, 1940: 83; 1953: 404–405; Barber, 1957: 44). The idea that the distribution of economic advantages is really secondary to recognition for one's contribution to society sounds vaguely like Marx's view that distributive issues will be relatively unimportant in the popular mood that prevails under true communism. Even more, however, this idea resembles Durkheim's image of society under the normal division of labour. In that hypothetical system, there will be an interplay of individual competition for rewards and group cooperation for collective ends. Thus, self-interest exists but is restrained and harnessed for the good of all. Although it involves some exaggeration, one could almost conclude that structural functionalists in North America see Durkheim's future society as an imminent occurrence. In their view, the normal division of labour, with equal opportunity and rewards based on merit and contribution, seems to have gone from a nineteenth-century hope to an emerging present-day reality (see Parsons, 1953: 433–439).

The Conception of Power in Structural Functionalism

So far, we have found that structural functionalists see social inequality as a stratified hierarchy of individual status-roles, ranked primarily by their value, in most people's minds, to society. Individuals compete for access to the higher status-roles because of the greater prestige they carry and, secondarily, because of the greater material and other rewards they offer. At least in democratic societies, this competition for rank is relatively open, because people have a reasonable opportunity to excel at what they do best. Such an arrangement is functional in that it ultimately serves both the individual's need to achieve and society's need to have vital positions filled by the most competent and qualified persons. In addition, the resulting stratification system also serves an *integrative* function, by mapping out where people fit in society and by providing a systematic pattern of norms for interaction with others (Parsons, 1940: 73).

But what is to ensure that people either accept where they fit in this structure or agree to the normative rules of the game that place them there? In particular, what is to stop certain individuals or groups from wresting away rewards belonging to others, usurping privileges not rightfully theirs, or otherwise forcing their will on the collectivity? This question really comes down to the role of power in social structures. In this final section, we will examine the approach taken by structural functionalists to the concept of power and the implications it has for their views on social inequality.

Power and the Problem of Social Order

Structural functionalists certainly discuss the idea of power, although their treatments tend to be rather brief and incidental to other issues (e.g., Alexander, 1987: 107; Johnson, 1993: 124). It is typical of structural functionalists to tie the concept of power directly to the fundamental question noted above: how do we maintain societal stability in the face of internal factions who might disrupt the social structure for their own ends? This so-called *problem of social order* is an especially important concern for Parsons, who consistently raises it in his own writings and who attributes a similar preoccupation to Durkheim as well (e.g., Parsons, 1937: 89, 307, 314–315, 402–403; 1951: 36–37, 118–119; see Giddens, 1971: 106).

The general conclusion reached by Parsons and other structural functionalists is that, apart from isolated incidents, the use of coercive power cannot be the means by which social order is attained, because force itself can only breed disruption and disorder in the end (e.g., Parsons, 1966: 246). Thus, to understand how stable societies exist, we must look for the source of social order elsewhere.

According to structural functionalism, social order arises mainly from legitimate and generally accepted bases of social control (e.g., Parsons, 1953: 418). A key process here is socialization, whereby most people learn and adopt a set of prescribed rules and norms that permit orderly, mutually beneficial social interaction. This point, of course, conforms with the structural-functionalist emphasis on popular consensus discussed earlier in this chapter. The crucial item to stress here is that these rules and norms are obeyed, in the structural-functionalist view, not out of fear of coercion or punishment by those in power, but primarily because the populace is instilled with the need or obligation to do what is right and to eschew what is not (Parsons, 1940: 74). Hence, the use of force in stable societies is rare because of the pronounced feeling among the people that both their rules and their rulers are essentially legitimate.

Power and Authority

This sense that legitimacy is the ultimate basis of social order is reflected in structural functionalism's conception of power. On the surface, most structural functionalists appear to employ Weber's classical definition of power: the capacity to exercise one's will, even in the face of opposition (e.g., Davis, 1949: 94–95; Johnson, 1960: 62; Parsons, 1966: 240; see Weber, 1922: 53). Typically, however, structural functionalists then draw a key distinction. *Power* does not in fact refer to all such instances of exercising one's will despite resistance, but only to those instances that are "illegitimate" or "not institutionally sanctioned." The term *authority* is reserved for those situations in which power is legitimate, institutionally recognized, or supported by "social consensus" (Parsons, 1940: 76; 1953: 391–392; 1966: 240, 249; Barber, 1957: 234; Williams, 1960: 96). Thus, for example, the government in a democracy exercises authority when it collects property tax, provided that taxation is legally under its jurisdiction; however, the government is exercising power if it seizes a person's property without recourse to legal statutes or some popular mandate.

This distinction between power and authority may seem reasonable enough. But, unfortunately, the tendency is for structural functionalists to believe that the concepts of power and authority exhaust all the situations in which some people hold sway over others. Thus, except for transitory cases of open, illegitimate coercion (power by their definition), most social relations are assumed to be based on legitimate influence (authority), since subordinates must harbour some degree of acceptance if they regularly obey their superiors. Yet such a broad view of what constitutes authority clearly subsumes a wide range of instances in which people obey others out of habit, custom, self-interest, a lack of real or perceived alternatives, and so on. While these situations do not entail openly coercive power, neither are they examples of truly legitimate control or authority (see Habermas, 1975: 96). Structural functionalists thus tend to blur important differences in both the intent and the meaning of the influences that operate in these cases. It is notable that this failure to acknowledge fully the wide range of situations in which neither power nor authority operates parallels a similar failure, noted earlier in this chapter, to appreciate completely the wide range of situations that involve neither open conflict nor general consensus.

Authority versus Domination

This tendency to emphasize the existence of authority, rather than other forms of influence, in social hierarchies is sometimes presented by structural functionalists as if it were consistent with Weberian theory. However, it is obvious from our discussion in Chapter 3 that this is not the case. The key difference between the structural-functionalist conception and Weber's analysis is that Weber does make an explicit distinction between authority, or genuinely *legitimate* domination, and other forms of domination based on habit, self-interest, and the like. The failure of structural functionalists and others to discern this crucial difference probably stems from Parsons, whose early translation of Weber's German works incorrectly treats domination (*Herrschaft*) and authority (*legitime Herrschaft*) as equivalent terms (see Weber, 1922: 62, 299; Giddens, 1971: 156; Alexander, 1983: 20-21).

These concerns over translation and terminology may seem like minor quibbles, but in fact they lead to fundamental difficulties in the way that structural functionalists conceive of power and domination in social hierarchies. Of course, we should remind ourselves again that we are dealing with a school of thought, and that not all writers in this school are the same. Nonetheless, the impression given by most structural-functionalist analyses is that virtually all enduring structures of domi-nation are basically legitimate. While this is a view that Weber would never espouse, it occurs rather consistently in structural-functionalist discussions. For example, Davis (1949: 95) asserts that "the whole social structure, the whole system of positions, may be viewed as a legitimate power system." The reasoning here is that "the line of power corresponds roughly with the hierarchy of prestige"—that is, with the popularly held ranking of what are the most important positions in society (Davis, 1949: 95). In other words, if the populace agrees that the most powerful posi-tions also have the highest evaluations of importance, then the power structure is legitimate by definition.

However, the crucial question that should be raised here is one we have already noted: who decides what positions are the most important and prestigious in society? The structural-functionalist assumption, as discussed earlier, is that society, or everybody in general, makes these decisions. If that were true, then the stratification system and its corre-sponding power structure would indeed have a legitimate and consensual basis in the popular will. But a plausible alternative is that it is largely the *people in power* who decide which positions are most important and thus most deserving of prestige and other rewards. In that event, the close correspondence among one's power, prestige, and privileges would have little to do with legitimacy or consensus but instead would flow mainly from the capacity of people in positions of domination to establish and maintain their own advantages, through force or other means.

This alternative image of power's role in social hierarchies is gen-erally similar to the views of Weber and Marx, which we reviewed in earlier chapters. It is apparent, then, that structural functionalism departs significantly from these classical perspectives. Although some structural functionalists concede that there are coercive and nonlegitimate aspects of power at times, there is little doubt that these factors are seen as of secondary importance for generating inequality. Power, in most cases, is not a *zero-sum* relationship, wherein people struggle for scarce resources and some win only if others lose. Rather, as modern social structures continue to develop and expand their mastery of the environment, there will really be more power for everyone, because of the increased opportunities to exploit and growing pool of resources to control (see especially Parsons, 1966; also Parsons, 1953: 436–437).

The Pluralism of Power

The last element in the structural-functionalist treatment of power that we should briefly discuss is its decidedly pluralist tone. As we have already seen, structural functionalists view modern society as a complex of structures and substructures, each of which performs important tasks for the overall system. This portrayal in itself implies a pluralist power structure, for it suggests numerous centres of jurisdiction, decision making, and control over resources (see Alexander, 1985: 9; 1987: 102). As well, each of these structures is in some sense another stratification system, with its own distinct distribution of powers and responsibilities to groups or individuals (e.g., Parsons, 1940: 86–87; Davis and Moore, 1945: 244). Thus, while we may speak of "the total 'power' system of a society," it is really made up of multiple components, "a plurality of other systems" that, when coordinated, ensure that societal problems are solved and that collective goals are achieved (Parsons, 1953: 388–389). Moreover, none of these structures by itself is seen as having "monolithic" or "paramount" control. The overall effect is a "separation of powers" so that no one interest group or elite has a "monopoly of influence" (Barber, 1957: 241–242; see also Parsons, 1953: 418, 426).

For individuals and subgroups in society, this pluralism means a great deal of opportunity for mobility into positions of value and authority. If, for example, an individual has little or no power in the political sphere, he or she nonetheless may attain high rank and influence in one or more of the economic, religious, or other substructures. Thus, the overall effect, both on the stratification system and the pattern of power, is considerable "openness," "looseness," and "dispersion" (Parsons, 1940: 86–87; 1953: 407, 430–432).

This portrayal of the modern power structure as fluid and open is of course entirely consistent with the other central tenets of the structural-functionalist perspective that we have outlined in this chapter. Equally clear are its departures from Marx's formulation. For Marx, the apparent dispersal of power in modern societies is largely a superstructural illusion, masking the underlying economic basis for all social domination.

The view of power among structural functionalists may seem vaguely similar to Weber's pluralist analysis. Even here, however, Weber's case for the multiple bases of power in society is different, because he sees the dispersion of power as more limited. Moreover, Weber retains a concern over the very real possibility that the domination of social life may eventually be centralized in a massive, omnipotent state bureaucracy. Structural functionalists do not seem to acknowledge the threat that state centralization poses for pluralism, at least with respect to capitalist liberal democracies such as the United States or Canada. It is only with respect to totalitarian societies such as the former Soviet Union or Nazi Germany that this threat is identified (Parsons, 1953: 407, 418; Barber, 1957: 241–242). It is probably no surprise that structural functionalism has achieved its greatest following and its best-known proponents in Western capitalist countries, especially the United States (Wrong, 1959). In these societies, the dominant belief system tends to accept social inequality as legitimate and normal as long as it results from equal opportunity, individual performance, and an absence of coercive power (Alexander, 1987: 102, 140).

Thus, we see once again that the structural-functionalist view of modern democratic societies seems much closer to Durkheim's vision of the future than it does to the predictions implicit in Marx or Weber. But is the moral, just, and normal division of labour an impending reality? The more skeptical among us may wonder just how close any society, including the United States or Canada, has come to achieving such a system. This is a question that we shall raise again.

SUMMARY

In this chapter we have taken an extensive look at the view of social inequality that has developed out of the so-called structural-functionalist school of sociological theory. In our discussion of modern structural functionalism, we reviewed its major concepts and some of the key criticisms that have been levelled against it. Our central concern, however, has been to examine the particular manner in which structural functionalists conceive of inequality, especially as it compares with the conceptions of Marx, Weber, and Durkheim. We have found that there are significant discrepancies between the structural-functionalist approach and each of these classical writers on most points. Among the crucial differences are the tendency for structural functionalists to see inequality mainly as a matter of individual stratification rank rather than class structure; to depart widely from the classical usage of class when they employ this concept; to perceive far more consensus, and far less conflict, in the processes that lead to inequality; to play down the importance of coercive power, compared with legitimate influence or authority, as the basis for inequality; and to see a significantly more open and equitable system of power and opportunity in present-day societies than is typical in earlier analyses.

The treatment of structural functionalism here has no doubt seemed much more critical and skeptical in tone than was evident in our assessments of other theories. This critical stance, however, should be partly tempered by the realization that we have had to examine the sometimes diverse school of thought that constitutes structural functionalism as if it were a single, consistent approach. Because of the need to summarize and distill the ideas of so many writers, some oversimplification has been inevitable. Thus, in some instances, the criticisms apply to certain writers more than to others and should not be seen as a categorical rejection of all the ideas and analysts identified with this school. This caution is important to note because of the tendency among some critics to present a mere caricature of structural functionalism and then to dismiss it entirely. The proposal here is to avoid an outright rejection of structural functionalism, at least until we have been able to examine the key issues again. Our final judgment of the structural-functionalist viewpoint, and of the views put forth by the earlier writers as well, should first take into account the most promising recent attempts to conceptualize the problem of social inequality. A review of these efforts is the task to be addressed in the next chapter.

CHAPTER SIX

Recent Perspectives on Social Inequality

> *All social interaction involves the use of power.*
> ANTHONY GIDDENS
> *A Contemporary Critique of Historical Materialism*, Volume 1, 1981
>
> *Exploitation without domination, or domination without exploitation, do not constitute class relations.*
> ERIK OLIN WRIGHT
> *Interrogating Inequality*, 1994
>
> *All systems of stratification are, by definition, systems of power inequity.*
> JANET CHAFETZ
> *Gender Equity*, 1990

INTRODUCTION

We have now examined four leading and long-standing perspectives on social inequality: the classical theories of Marx, Weber, and Durkheim, and the more recent structural-functionalist explanation. These early views, of course, do not exhaust the theoretical approaches to inequality that we could consider. In particular, numerous formulations have come to light in more modern times. Clearly, it is not possible to review and assess all of these in detail. It can be argued, though, that such a thoroughgoing review is unnecessary, provided that we can arrive at a set of contemporary theorists who, taken together, incorporate most of the major conceptual advances to be found in the larger group of writers.

The central goal of this chapter is to conduct a selective assessment of the most promising or influential of the more recent perspectives on social inequality. The chapter begins with an analysis of six contemporary writers whose works, in varying degrees, can be seen to contribute to a *general* theory of social inequality. These six approaches offer perspectives that can be useful for conceptualizing many types of inequality, including not only class, for example, but also gender, race, ethnicity, and so on. The six writers to receive special attention are: Ralf Dahrendorf and Gerhard Lenski, whose works probably rank as the best-known departures from structural functionalism to come out of the 1950s and 1960s; Nicos Poulantzas and Erik Olin Wright, who are among the foremost Marxist thinkers to achieve distinction in the study of inequality in recent decades; and finally Frank Parkin and Anthony Giddens, whose efforts represent what are arguably the leading non-Marxist or neo-Weberian approaches of the contemporary period.

Inevitably, the selection of these six writers over others is to some extent not a matter on which complete consensus is possible. Any attempt at a review of this sort means that some theorists of note will necessarily be excluded from detailed analysis. For example, we do not consider earlier Marxists such as Lenin (1917) or C. Wright Mills (1951, 1956), or more recent writers who have been influenced by Marx, such as Immanuel Wallerstein (1974, 1979, 1989), Claus Offe (1974, 1984), and Jurgen Habermas (1975, 1984), among others. In addition, a whole range of broadly Weberian or non-Marxist theorists is not given detailed consideration (e.g., Blau, 1964, 1977; Lukes, 1974, 1978; Wrong, 1979; Turner, 1984; Collins, 1985, 1986a, 1986b, 1988b; Mann, 1986, 1993; Goldthorpe, 1987; Runciman, 1989; Blalock, 1989, 1991; Goldthorpe and Erikson, 1992; Scott, 1996; Tilly, 1998). The view taken here is that the key insights of these and other writers are, for our purposes, largely represented in the works of the six theorists who have been chosen for review. On certain points, of course, reference to numerous other theorists will be made where it is appropriate to do so.

In addition to examining the six theorists noted above, this chapter will also consider some leading examples of more specialized perspectives on inequality. Of particular interest are those that focus on gender and race. In the past several decades, we have seen the emergence (or more accurately, the re-emergence) of these two issues as prominent areas of theory and research in social inequality. In fact, some observers have expressed concern that the renewed interest in gender and race has become so pronounced that other central issues in the field of inequality, most notably the study of class, may have been set aside or treated as only secondary problems (e.g., Wright, 1985: 57). One goal of this chapter is to argue that, although the increasing interest in gender, race, and other nonclass forms of inequality is clearly justified, there is also no question that class continues to be a core concept in any general theory or explanation of social inequality in modern times.

Space limitations preclude a comprehensive examination of the extensive theoretical work that has been generated on race and especially on gender in recent years. Nevertheless, several of these more specialized perspectives are of note here, not only because of their contributions to the study of gender or racial inequality, but also because of their significant affinities and parallels with the best general approaches. It will be argued later, in fact, that the overall portrait of social inequality that emerges from the more general perspectives considered in this chapter is largely consistent with, and often corresponds closely to, many leading perspectives that have arisen in the analysis of gender and race.

One final issue that will be assessed at the close of Chapter 6 concerns the growing interest in understanding the structure of social inequality at the international or global level. It is beyond our purposes to conduct a detailed assessment of this question. This is in part because the view taken here is that the major insights to be found in the leading past and present theories of social inequality are, for the most part, sufficient for conceptualizing the nature of inequality, not just *within* individual societies, but *among* the various nations of the world. This is not to say that there are no new insights to be found in the work of the writers who have specialized in the study of global inequality. However, our brief review will seek to demonstrate how this work, in many ways, is quite consistent with, and at times builds upon, the more general perspectives provided by several of the classical and contemporary perspectives that are considered in this volume.

Throughout the chapter, the reader should remain alert to the key areas of similarity and dissimilarity, not only among the selected recent approaches but also between each of them and the classical views we have already examined. Obviously, our discussion would be much too involved were we to deal with these comparisons in all their minute aspects. Hence, rather than a comprehensive exposition of all the recent conceptual schemes, we shall conduct a more limited investigation, one that critically evaluates each approach but focuses primarily on the four major topics that were outlined in the opening chapter. These four points include each author's stance on the concept of class or class structure; the significance of power for explaining structured inequality generally in society; the interconnection of the state with the economic and other structures of society; and the prospects for reducing or ending social inequality in the future, through socialist revolution or other means.

The recent perspectives to be considered in this chapter, combined with the classical views already discussed, occupy a wide theoretical spectrum that, if analyzed in detail, reveals countless subtle differences in conception and emphasis. Yet, when these perspectives are examined on a more general level of analysis, one can argue that their similarities are almost as striking as their differences. The approach taken here is to adopt a middle ground, acknowledging both the essential distinctions and interesting similarities among perspectives, while also simplifying and clarifying the presentation whenever possible. As a step in this direction, it is instructive to array the various theorists along a general continuum between two poles that, for want of more precise terms, are labelled *left* and *right*. To take the most extreme illustration, those at the left pole see social inequality arising purely out of conflict or struggle between antagonistic groups; trace the root of inequality to the single factor of class location or control of economic power; and stress the major social problems that inequality engenders, problems that can be alleviated only through radical social change. In contrast, those at the right pole see social inequality wholly as the result of consensual interaction among individuals competing with one another according to agreed-upon rules of conduct; believe that inequality flows from an extreme plurality of factors, not from one or a few sources; and stress the positive effects of inequality, especially for societal stability and integration.

Although no theorist under scrutiny here precisely fits either of these two extreme images, we can estimate each one's *relative* position on the continuum between the poles, as shown in Figure 6.1. Note that the location of each perspective on the left–right scale is cross-referenced with the approximate time of its emergence or prominence, especially in

Figure 6.1

The Chronology of Major Theories of Inequality and Approximate Locations on Loosely Defined Left–Right Continuum

Period of Emergence or Prominence

Before 1900	Marx, early Marxism			
1900 to 1920			Weber Durkheim	
1920 to 1950				Structural Functionalism
1950 to 1960			Dahrendorf	
1960 to 1970		Lenski		
1970 to 1990	Poulantzas Wright	Giddens Parkin		
Since 1990	Theories of gender and race			

Left——————————————————————————————**Right**

(Inequality based on struggle, rooted in class or economic power, must be changed by radical action)

(Inequality based on consensus, has extremely pluralist roots, provides stability and other benefits for society)

North American sociological circles. This procedure produces an interesting pattern, one that roughly resembles a "U" laid on its side.

We should probably avoid grand explanations for this pattern, but one way to make some sense of it is to consider the manner in which leading intellectuals respond to the climate of the times in different historical periods. We have already discussed how Weber's work in many respects was a critical reaction to Marx, or at least to the manner in which Marx's main premises were represented by the subsequent groundswell of Marxist thinkers in the late nineteenth and early twentieth centuries. Weber's critique, which is paralleled in Durkheim's writings to some extent, was meant to be a positive, or constructive, one. However, it seems to have set in motion a much wider swing away from the left pole than Weber intended. This swing was accelerated significantly by Parsons's sometimes peculiar interpretations of Durkheim and Weber, which eliminated from these writers' works most of the affinities with Marx that we have outlined in previous chapters. The culmination of this trend was the emergence of the structural-functionalist school as the dominant force in sociological thought by the 1950s, especially in the United States.

At about this time, as was noted in Chapter 5, certain influential writers came to react against the inaccuracies that they perceived in the more extreme versions of structural functionalism. Dahrendorf was one of the leading critics of what was seen as a false, utopian representation of societal harmony, stability, and consensus by the structural-functionalist school (e.g., Dahrendorf, 1958, 1959, 1968; see also Mills, 1959; Wrong, 1961). The work of Dahrendorf and others provided a catalyst that sent the pendulum swinging back again, away from the right end of the continuum and into a broad middle range of rather diverse viewpoints. Lenski is among the most notable here as someone who himself moved from a structural-functionalist stance to a "synthesizing," partly Weberian position in the 1960s (Lenski, 1966: 435).

Although they differ in other respects, writers like Dahrendorf and Lenski share a desire to combine in one perspective key elements from both the left and the right poles, from Marx or Marxism on the one hand and from structural functionalism on the other. At least two important developments have followed in the wake of such efforts. First of all, the changes they have stimulated seem to have come more at the expense of structural functionalism than of Marx. In other words, whereas Marx's ideas have had a continuing influence to the present day, structural functionalism has virtually abandoned the topic of social inequality, at least since the 1960s. This abandonment may be a tacit admission by structural functionalists of fundamental difficulties in their approach to inequality, or it may reflect the simple truth that, after all, the subject is of secondary concern to them and not deserving of sustained attention. Whatever the reason, structural functionalism has been largely supplanted, at least at present, as a serious alternative to Marxism in the current literature on social inequality.

The second important consequence of Dahrendorf's and Lenski's work has been a revival of Weberian thought. As we shall see, the emphasis that these two theorists place on power, combined with their pluralist conceptions of class and power structures, returns us to some version of the Weberian perspective. Thus, the pendulum moves back to Weber and the renewal of an old rivalry: the theoretical debate between Weber and Marx. In fact, some writers have argued that, even in the current era, the trend of thought in the study of social inequality oscillates between these two writers, that "the all-pervasive influence of Marx and Weber . . . is, if anything, more pronounced today than at any other stage" (Parkin, 1978: 601; see also Collins, 1986a: xi; Wiley, 1987: 25; Sayer, 1991: 5; Scott, 1996: 3–5, 17). In many respects, this is the same argument that will be put forth in this chapter. The ideas of Marx and Weber, especially as they have been refurbished by the debates exemplified in the works of such writers as Poulantzas, Wright, Parkin, and Giddens, provide most of the conceptual elements that are required for examining the general problem of inequality in modern societies. Wright and Giddens are particularly noteworthy for their sustained efforts to rethink and reconstitute their theoretical approaches to inequality and related issues (e.g., Wright, Levine, and Sober, 1992; Wright, 1994, 1997, 2005; Giddens, 1994, 1998, 2000a, 2000b).

Because of their relative emphases, it has been typical to classify Poulantzas and Wright as neo-Marxists, with Parkin and Giddens usually viewed as neo-Weberians. Such a classification can be helpful, provided it is applied cautiously and at a general level of discussion. Thus, it is fair to refer to Poulantzas and Wright as neo-Marxists because they provide us with newer, revitalized analyses based on Marx's first principles. At the same time, though, we should also note that such categories exhibit wide internal variations. This is well illustrated by Parkin, who readily embraces the neo-Weberian label but disagrees with several other writers to whom he gives the same title (Parkin, 1972: 29–33; Parkin, 1979: 112). But probably the greatest difficulty with these classifications is that they cannot deal precisely with those, like Giddens, who prefer to be allied with neither the Marxist nor the Weberian camp (Giddens, 1981a: 296–297; 1981b: 1; 1984: xxxvi). Although Giddens acknowledges the contributions to his approach from both Marx and Weber, it is true that some of the more subtle elements of his work do not stem from either of these classical thinkers. And yet, at a broader level of discussion, there is justification for locating Giddens relatively near the neo-Weberian circle. As we shall see, Giddens, like Weber, is greatly concerned with offering a critical but constructive appraisal of Marx's historical materialism (see especially Giddens, 1981b, 1985). Like Weber, Giddens places considerable stress on the complexity of social inequality and the various forms it can take. Finally, like Weber, Giddens uses as his conceptual centrepiece the idea of power and its role in the structures of domination that shape social interaction in modern societies. For these reasons and others, attaching the neo-Weberian label to Giddens is in some respects justifiable.

Figure 6.1 (page 117) also depicts the more prominent role played by gender and race in social inequality in more recent work. As we have already noted, problems of gender and race were important issues in social theory before the current period. However, these two areas of inquiry, in a sense, have been rediscovered of late, and are now the major issues of interest to many contemporary social theorists. While it is difficult to single out individual writers in the vast array of current gender theorists, the works of people such as Rae Blumberg (e.g., 1978, 1984, 1989, 1991a, 1991b, 2004; see also Blumberg et al., 1995), Janet Chafetz (e.g., 1984, 1988, 1990, 1997, 1999), R.W. Connell (1987), Margaret Andersen and Patricia Collins (1995), and Sylvia Walby (e.g., 1990, 1997, 1999, 2002, 2005) provide some of the best examples of leading discussions of gender inequality in the present day (see also Wallace, 1989; England, 1993; and others). Similarly, the ideas of writers such as William Wilson (1973, 1987), Edna Bonacich (1972, 1985), Susan Olzak (1992), and Joane Nagel (e.g., Olzak and Nagel, 1986) are among the most fruitful for understanding racial or ethnic inequality in contemporary societies (see also Bonilla-Silva, 1997; Satzewich, 1998; Henry et al., 2000).

In Figure 6.1, the sets of perspectives dealing with gender and race/ethnicity are located in roughly the same area on the left–right continuum as that occupied by the four recent Marxist and non-Marxist writers we have selected. This positioning is consistent with the "left-of-centre" views of inequality taken by most leading writers dealing with gender and race in the contemporary period. The location also signifies a central contention of the current analysis, which is that although the best present-day perspectives on gender and race have their own distinct contributions to make to the study of inequality, they share in common some of the same key themes and concerns found in the debates between contemporary Marxist and Weberian theorists. In particular, they all address, in various ways, two core issues: first, how race and gender inequality intersect or overlap with class inequality; and, second, how these and other types of inequality may ultimately be tied to differences in power or domination.

As we shall see, this common ground is perhaps most evident in the more recent works of Wright and Giddens. These two writers are notable, as well, for acknowledging that some of their earlier analyses were insufficiently attentive to the special conceptual problems posed by gender and race, and for seeking to include more of these key concerns in their subsequent writings (e.g., Wright, Levine, and Sober, 1992;

Wright, 1994, 1997, 2005; Giddens, 1989, 1990, 1992, 1994, 2000b, 2002, 2003; see also Giddens and Diamond, 2005).

We are now in a position to review the recent perspectives. We begin chronologically with the work of Dahrendorf and then Lenski. The neo-Marxist views of Poulantzas and Wright will be considered next, followed by Parkin's neo-Weberian critique of their general position. Next comes Giddens's formulation, which in some respects may be the most inclusive of the more recent general perspectives. We then consider a selection of perspectives on gender and racial inequality and assess their potential for extending and supplementing the best insights from the more general approaches. Finally, we conclude the chapter with a brief overview of approaches to the study of social inequality at the global level of analysis.

RALF DAHRENDORF

Class Conflict in Industrial Society

In *Class and Class Conflict in Industrial Society* (1959), Ralf Dahrendorf presents his first detailed formulation of the problem of inequality in modern societies. While his later writings also address this same issue, reaffirming many of his original ideas and modifying others, his early work still stands as his most influential and cogent theoretical statement on the nature of social inequality (see also Dahrendorf, 1979, 1988, 1990a, 1990b, 1997, 2004).

A central claim in Dahrendorf's analysis is that neither structural functionalism nor Marxism is adequate by itself as a perspective on society, since the former pays too little attention to the realities of social conflict, while the latter ignores the obvious evidence of consensus and integration in modern social structures (1959: 122–124, 158–160). As an alternative, Dahrendorf proposes to "draw from Marx what is still useful" and to incorporate it with certain promising elements of the structural-functionalist viewpoint (1959: 118). Hence Dahrendorf contends, following Marx, that conflict is still a basic fact of social life and can have positive consequences for society as a spur to progressive social change. In this way, Dahrendorf seeks to dissociate himself from those structural functionalists and others who saw an "end of ideology" during the 1950s, who proclaimed a complete end to the conflict of ideas and class interests that preoccupied Marx in the nineteenth century (e.g., Riesman, Glazer, and Denney, 1953; Bell, 1960; for discussion, see Giddens, 1973; Benson, 1978). Dahrendorf's more recent work sustains his essential argument that while we can identify "a wide consensus" in many democratic societies, there is also no question that "conflict is the great stimulus of change" and is still crucial as "a potential for progress" (Dahrendorf, 1997: 28; 1988: xii, 111).

However, despite Dahrendorf's long-standing belief in the fruitful consequences of social conflict, his conceptual affinities with Marx remain rather minor in the end and, on some important issues, do not distinguish him clearly from some of the structural functionalists he criticizes. In fact, Dahrendorf shares with structural functionalism several fundamental ideas, including a general optimism and faith in the political and economic institutions that arose from the prosperity and stability of the period following World War II. Dahrendorf perceives several important changes and improvements in the capitalism of Marx's age that have led to a new social structure called "industrial" or "postcapitalist" society (1959: 40–41). In general, postcapitalism involves a much more complex system of inequality than can be captured by the simple split between capitalists and workers, the "two great and homogeneous hostile camps with which Marx was concerned" (1959: 48). Postcapitalism is marked, first, by a diverse class structure and, second, by a very fluid system of power relations. Moreover, it is a society in which the resolution of class conflict has been "institutionalized"—legitimately incorporated within the state and economic spheres—so that the drastic class strife of Marx's time has been made obsolete. Let us now consider each of these points and their relevance to Dahrendorf's views on the future of inequality in postcapitalist society.

Ralf Dahrendorf,
born 1929

Getty Images.

Class Structure in Postcapitalism

Dahrendorf rejects Marx's dichotomous, two-class system because it is too simplistic to be applicable in postcapitalist society and because its stress on property ownership as the single distinguishing class characteristic has become outdated. For one thing, the capitalist class has been "decomposed" by the rise of the "joint-stock company," which separates simple ownership from actual control of economic production. Although Marx was certainly aware of this new form of business organization, owned by stockholders and run by managers or corporate executives, Dahrendorf claims that Marx underestimated how much power this arrangement would take from owners and give to executives, who may own no part of the enterprise but still make the crucial business decisions (1959: 47).

A second important complication in modern class structures concerns diversification within the working class itself. Like Weber, Dahrendorf criticizes Marx's view that the proletariat will eventually become a homogeneous collection of relatively unskilled machine operators. On the contrary, there are now increasingly elaborate distinctions among workers regarding skill levels, life chances, and prestige, with a need for more, not less, skilled personnel to run and maintain the sophisticated machinery of industry. The result is "a plurality of status and skill groups whose interests often diverge" (1959: 51, 277; 1988: 77).

The third major factor in this pluralism of classes involves those categories of people who are neither bourgeois nor proletarian precisely but are lumped by various writers under the title "new middle class." To Dahrendorf, some of these people, including salaried white-collar employees, overlap partly with the old working class, while others, including some bureaucrats and the executives noted earlier, may have vague affinities with the old capitalist class. In both cases, however, these intermediate groupings are distinguishable in key respects from both the bourgeoisie and the proletariat, confirming once again that "the pleasing simplicity of Marx's view has become a nonsensical construction" (1959: 57; 1988: 77).

For these reasons and others, Dahrendorf contends that Marx's idea of class can be salvaged only if its entire meaning and definition are altered to reflect the changes in modern social structures. Dahrendorf thus makes an alternative proposal. Because of the separation of ownership and control, the growth of corporate and state bureaucracies, and other manifestations of complex organizational hierarchies, it is apparent that the crux of social inequality is no longer the antagonistic property relations between capital and labour, but the *authority* relations arising within a wide variety of social and organizational settings (1959: 136–138). Authority relations are crucial because, while it is possible to imagine an end of inequality in property holdings or in such advantages as income and prestige, it is impossible to conceive of social organization without inequalities in authority (1959: 64, 70–71, 219). Moreover, "authority is the more general social relation," with property "but one of its numerous types" (1959: 137).

According to Dahrendorf, then, any useful definition of class in postcapitalist society should include the key idea of authority, and the term class relations should refer not just to the conflict between economic groupings but to *all* situations of struggle between those who have authority and those who do not. Hence, Dahrendorf explicitly defines classes as "social conflict groups," distinguished from one another by their "participation in or exclusion from the exercise of authority" (1959: 138, 247; 1988: 112). These conflicts are most telling within the major institutions and organizations, especially in the economic or industrial sphere and in the political structure or state. In theory, however, they can occur in any hierarchical authority structure, any "imperatively coordinated association" (1959: 138–139).

Dahrendorf's redefinition of class in this manner is intriguing and, on the surface, may appear to have merit. His claims that inequalities in authority are inevitable, and that property relations are but one manifestation of such inequalities, seem difficult to deny. Indeed, most analysts, including Marx and Engels, acknowledge both of these points from time to time (e.g., Marx, 1867: 300–331; Engels, 1872; 1890b: 463–465).

But does the recognition that authority is an important factor in social relations justify its equation with the concept of class? The answer to this question must be no. Dahrendorf himself concedes that his proposal for defining authority relations as class relations is "pragmatic" and "reversible" (1959: 201). Three decades after writing *Class and Class Conflict in Industrial Society*, he concedes again that he defines class "without undue claims for certainty" and in a manner that may seem "cavalier" (1988: 47, 112). However, such temporary or tentative attempts at defining class make it difficult to arrive at a conception of class that is sufficiently clear and consistent for the cumulative understanding of theory in this area. Dahrendorf's class definition is also at odds with the classical views of Marx and Weber in many instances, as well as with established usage. In addition, his definition has some inherent logical problems. For example, if every conflict over authority is a class relation, then there are an infinite number of classes, which largely robs class of its meaning and value as a sociological concept (see Dahrendorf, 1988: 47). As others have pointed out, Dahrendorf's conception also leads to certain problematic conclusions because every confrontation over authority, even between parent and child, for example, is by definition a class conflict (e.g., Giddens, 1973: 73). Such difficulties are somewhat surprising, since Dahrendorf is sometimes critical of theorists who confuse the general with the specific; yet, by seeing all authority relations as class relations, instead of viewing class relations as really one crucial type of authority (or power) relation, Dahrendorf appears to commit a similar error himself (1959: 137).

Dahrendorf on Power and Authority

It should be clear by now that Dahrendorf's general conceptions of class and inequality are closely tied to his view of authority or power. This emphasis is reminiscent of Weber, with whom Dahrendorf shares certain broad affinities. These include his pluralist view of class and power structures; his belief that authority hierarchies are inevitable in all advanced societies, capitalist or otherwise; and his interest in the growing concentration of influence within formal bureaucracies, especially in politics and industry (1959: 299; 1988: 25–26).

It is only when we examine Dahrendorf's actual definitions of power and authority that his differences with Weber become apparent. One curious aspect of Dahrendorf's discussion is that he claims to adopt Weber's view of power and authority but, in the end, uses a terminology that is actually closer to Parsons and the structural-functionalist school. This inconsistency may be most serious in his treatment of Weber's idea of domination (*Herrschaft*). Like Parsons, Dahrendorf generally treats all situations of domination, of patterned or structured power relationships, as if they were legitimate (1959: 166). As a result, he does not distinguish clearly between authority (truly legitimate power) and those cases where subordinates give regular obedience to superiors, not out of a belief in the justice or legitimacy of the relationship but for a variety of other reasons. At times, Dahrendorf seems to acknowledge the difference between domination and authority, but generally he blurs this distinction and so creates some confusion (1959: 176; 1979: 117). Dahrendorf's subsequent writings reveal a more clearly Weberian position on the idea of power, with fewer functionalist overtones than are found in his original conception (1988: 26–27). Even so, he does not provide an explicit definition of either power or domination in his current work, so that it is difficult to judge exactly how his use of these concepts has changed, except to note that he tends now to avoid their equation with the idea of authority.

A final difficulty with Dahrendorf's approach to power is his contention that all situations of conflict must involve two and only two contending parties (1959: 126). This means that all class conflicts, and all authority or power relations, must involve a "dichotomy of positions" between those who possess power and those who are deprived of it (1959: 170, 238). Such a view obviously departs from Weber, who saw power as a graded phenomenon, varying in degree within bureaucracies and other hierarchies. Dahrendorf's dichotomous view also seems too simple for understanding or explaining power relations in real societies. It is possible to identify many illustrations that run counter to such a perspective. The complex authority network of a modern corporation, comprising owners, top executives, middle managers, and salaried employees, among others, cannot be reduced to a set of dichotomous relations. The numerous examples of multiparty political systems, as in Canada or Great Britain, provide further evidence against the claim that confrontations over power always involve only two factions. Dahrendorf's dichotomous conception of power relations also seems to be at odds with his own generally pluralist view of social structures and his occasional references to "gradations" of power or authority in society (1959: 196; 1988: 77).

The Institutionalization of Class Conflict

Unlike some of Marx's more extreme critics, Dahrendorf readily accepts that conflict is inherent to social structures and treats it as a definitive feature of all class and power relations. However, what distinguishes him from Marx, and what explains his location toward the structural-functionalist end of our left–right continuum, is Dahrendorf's special conception of modern class conflict. Although he agrees that Marx's "old class conflict is by no means fully played out," he also contends that today's class conflicts are "far removed from the ruthless and absolute class struggle visualized by Marx" (1988: 45; 1959: 66). Rather than violent class warfare, the contending parties in postcapitalism usually engage in regulated, or "institutionalized," conflict. In other words, they have "agreed on certain rules of the game and created institutions which [have] provided a framework for the routinization of the process of conflict" (1959: 65, 225–227). Dahrendorf maintains the strong conviction that revolutionary conflicts are largely lamentable events, or "melancholy moments of history" (1988: 1, 194). His most recent writings also make it clear that, for Dahrendorf, "to be fruitful, conflict has to be domesticated by institutions" and must be guided by "rules" and a regard for "liberty"; this enables progressive change to occur, but in a way that is "strategic" and reformist, not revolutionary (1988: xii, 35, 107, 174, 193; 1997: 28, 42, 48; 2004: 22; see also 1990a, 1990b).

The institutionalization of class struggle is apparent within the economy, as well as within the various legal and political organs of the state. In the economic sphere, it is best illustrated by the spread of unionization and collective bargaining, whereby labour and management pursue their conflicting interests and resolve disputes through conciliation, mediation, or arbitration. In the legal–political sphere, institutionalized conflict is evident in the settlement of grievances through the law courts and the negotiation of legislation and policy decisions through parliamentary debate (1959: 66, 228–231). In this form, conflict is an essential force for social change, as Marx believed, but it also is an important source of coherence and unity, because of its problem-solving function (1959: 206–208).

Thus, even though Dahrendorf speaks of conflict almost as much as Marx does, Dahrendorf's stress on the regulated nature and integrative function of conflict reveals that his affinities with Marx are more tenuous than his ties to certain versions of structural functionalism. Dahrendorf's closer affinities with structural functionalism are reflected in his very definition of conflict, which is so broad that it could include everything from outright war to friendly competition, even sports contests and games (1959: 135). This diluted view of conflict, combined with Dahrendorf's implication that class and power relations are all somehow rooted in authority structures, leads his portrayal of social inequality in a different direction from that of Marx. Social inequality under postcapitalism is a product of the contest for advantage among conflicting interests; however, this contest ideally occurs according to institutionalized regulations within both the economy and the state, and under conditions of considerable pluralism and legitimacy. The conflict or struggle that arises is more a means for keeping contemporary industrial society healthy and progressive than a cataclysmic force for the revolutionary overthrow of capitalism (1959: 134). More recently, Dahrendorf still takes the view of a "radical liberal," whose preferred version of society is one that is open to creative and frequently conflict-driven patterns of change and reform, but that is also guided in these activities by constitutional regulation and "the rule of law" (1990a: 508, 525; 1990b: 71, 80, 122; 1997: 28, 42; 2004: 22).

The Future of Social Inequality

Given Dahrendorf's stress on the constructive and progressive nature of modern social struggle, it is not surprising that he has maintained a generally optimistic view of the future of inequality in postcapitalist society. Some of the projections in his early work on this question suggest that class conflict should be increasingly institutionalized in the future, and that equalities of opportunity and condition should themselves become part of the established institutions of society. Echoing Durkheim's views on the future to some extent, Dahrendorf sees the state as playing a key role here through the implementation of public education to increase the mobility chances of the lower strata. Although complete equality of educational opportunity might not yet be achieved, Dahrendorf argued in his early writings that there were sufficient advances in this regard to expect the gradual weakening of class barriers. His view was that "no social stratum, group, or class can remain completely stable for more than one generation" (1959: 59). He also saw signs that the state was helping to narrow the inequalities between top and bottom, by establishing social-welfare programs and tax laws that redistribute wealth from the rich to the poor. Thus, Dahrendorf initially envisioned a continued reduction of inequality over time, contending that "the process of levelling social differences cannot be denied" (1959: 63, 274).

Since that optimistic interlude of the 1950s, many analysts have come to question whether the promised reduction of inequality would ever come true. As recently as the late 1970s, Dahrendorf was among the few observers to claim that the drive for greater equality had *not* weakened (1979: 128). In fact, although he expressed support for increasing equality in society, he also feared that too much equality could pose problems. In Dahrendorf's opinion, total equality carries with it many dangers, "for a society in which all are equal in all respects is one devoid of realistic hope and thus of incentives for progress" (1979: 123; see also 1997: 63–65).

More recently, Dahrendorf's position has changed, for he now agrees with most other analysts that inequalities have probably become more pronounced in the contemporary period, even in such affluent nations as the United States and the countries of western Europe (1988: x, 149–150). He also has expressed concerns about the loss of cohesion and threats to freedom and opportunity that may occur if an increasingly competitive world allows these inequalities to grow still further (Dahrendorf, 1995; 1997: 66; 2004: 21–23). Nonetheless, his prevailing view has generally been that, in the developed societies at least, "the overwhelming majority of the people have found a reasonably cosy existence" (1988: 154). Moreover, while he agrees that a more equitable distribution of material wealth and "provisions" is now a higher priority than it was several decades ago, he remains consistent in arguing for the necessity of individual freedom and the hope for personal improvement that such freedom entails, even if it means some inequality of condition among citizens (1988: ix, xi, xiv, 8, 18, 73; 1997: 48, 62).

On this key point, then, Dahrendorf takes a position that is once again closer to structural functionalism than to Marxism. He sees social inequality as necessary to motivate individuals to pursue excellence in a free and open society. He also consistently rejects any attempt to establish totalitarian, state-socialist societies (1959: 318), although he now believes such systems are essentially "dead" in any case (1990a: 508). To Dahrendorf, socialist systems try to impose equality through a grey sameness that stifles progress and freedom. He fervently believes that hope and liberty are indispensable to any just society. However, as long as all citizens are equal in their basic rights and opportunities, regardless of their racial, religious, gender, and other differences, then it is clear to Dahrendorf that "hope springs from difference rather than sameness, and liberty from inequality rather than equality" (1979: 140; see also 1988: ix, 8; 1990a: 36; 1997: 58, 62–64).

Summary Observations

In this section, we have reviewed and evaluated the major elements in Dahrendorf's discussion of class conflict in industrial society, with special reference to his views on social inequality. It is apparent that there are some difficulties in Dahrendorf's conceptions of class, power, and conflict, at least in his earlier writings. It is also the case that some of his early claims and assumptions suggesting a declining level of inequality in modern times have had to be revised in the light of more recent evidence and analysis. Nevertheless, Dahrendorf's work is notable as perhaps the first major attempt to move away from a doctrinaire structural-functionalist perspective, by incorporating aspects of Marx and Weber into an alternative orientation to social inequality. Much of his work also provides one of the most eloquent arguments in favour of citizenship rights and individual freedom among sociologists writing in the contemporary period. The overall result takes us just part of the way to a more balanced or comprehensive perspective on social inequality and, apart from some recent shifts to a more clearly Weberian treatment of power, does not relinquish some basic structural-functionalist leanings. However, Dahrendorf's formulation, especially his original thesis on class and class conflict in the modern era, continues to be one of the most important stimuli for many of the approaches that we shall examine in the remainder of this chapter.

GERHARD LENSKI
Social Inequality as Power and Privilege

The second important theoretical perspective to be considered in the more recent literature on social inequality is Gerhard Lenski's analysis, *Power and Privilege* (Lenski, 1966). Rather like Dahrendorf before him, Lenski proceeds on the assumption that a comprehensive approach to social inequality must take into account the entire spectrum of views between "radicals" such as Marx on the one hand and "conservatives" like the structural functionalists on the other (see also Wenger, 1987: 45). He portrays his own formulation as an initial step toward a "synthesis of the valid insights of both the conservative and radical traditions," a stance that he readily acknowledges brings him very close to Weber (1966: 17–18). The final result is a position roughly midway between the poles of our left–right continuum.

Lenski's proposed synthesis involves a selection of certain basic premises from both the left and the right, moderated and amalgamated in a generally pragmatic fashion. His overall conclusion is that some degree of social inequality is inevitable, as conservatives argue, in part because humans differ in their effort, strength, intelligence, and so on, and tend to use these favoured traits to further their own interests or those of special groups to which they belong. This does not mean that all inequalities result from differences in "natural endowment" (1966: 32). In fact, most do not. Neither does this mean that people are incapable of the altruism

Gerhard Lenski, born 1924

Reprinted by permission of Gerhard Lenski.

and collective orientation that Marx and other radicals have stressed. Nevertheless, partisan interests do normally receive first priority because of tendencies in human nature, and a good deal of the apparent co-operation and selflessness that radicals point to occurs more from "enlightened self-interest" or necessity than from a universalistic orientation to the common good (1966: 25–32, 441–442; 1988: 169–170; see also 2005: 48–50, 216–218).

Where Lenski differs markedly from the conservatives is on their insistence that such selfish tendencies in human nature can somehow produce just and legitimate social hierarchies. Thus, Lenski rejects the structural-functionalist view that most people agree on the value or prestige attached to positions in society and that this common evaluation gives rise to corresponding differences in both power and material rewards. On the contrary, Lenski argues the reverse view: that differential access to power is ultimately what determines inequalities in material privilege and that power and privilege together determine most of the prestige attached to various groups or individuals in society (1966: 44–46). On this crucial point, then, Lenski believes that radicals like Marx are nearer to the truth. The radical perspective recognizes that the outward stability of social structures may say little about the extent to which such structures are endorsed by the populace and may instead mask fundamental antagonisms between dominant and subordinate factions. Of course, some level of genuine legitimacy is likely to be accorded the power structure, especially by citizens of Western democratic countries. Even here, however, legitimacy is often far from complete. Echoing Weber, Lenski suggests that the general acceptance of existing social hierarchies seems to result from at least two additional influences: the potential coercive force of the dominant group; and the "inertia" of custom, habit, and conventional beliefs, which holds back the impetus for struggle and change (1966: 32–35, 41).

It is clear, then, that Lenski considers differences in power to be the pivotal factor responsible for the structure of inequality in society. Because of this emphasis, it is preferable to begin with a review of Lenski's conception of power and then trace its influence on his discussion of class, the role of the state, and, finally, the prospects for social inequality in the future.

Power and the Multidimensional View of Inequality

It is immediately apparent that Lenski adopts a largely Weberian conception of power, though he reveals certain differences with Weber that should also be noted. Like Weber, he begins by defining power as the capacity to carry out one's will despite opposition. Following Weber, he also maintains that coercive force is "the most effective form of power." Of course, since any enduring system of domination must minimize violent upheaval, it is important that force not be used exclusively or imprudently but should "recede into the background to be used only when other techniques fail." Nevertheless, Lenski agrees with Weber's position that force is the "ultimate guarantee" and "the foundation of any system of inequality" (Lenski, 1966: 50–51).

Eventually, the factions that control the use of force try to have their power formally established and officially recognized, by creating or rewriting laws to protect their general interests. In modern democracies, such laws are not blatantly self-serving in most cases but, on the contrary, are frequently couched in universal terms that give them an air of popular legitimacy. Typically, for example, laws not only spell out the powers of those who rule, but also specify the formal rights of subordinates. Thus, a small minority may control the political structure, but there are provisions for the mass of the population to voice their wishes or their grievances through regular elections, referenda, or other means. Still, despite this apparent two-way flow of power between the top and the bottom, the relationship is decidedly asymmetrical, with power running more strongly and with far greater effect from the dominant to the subordinate faction (1966: 52–53, 58).

Lenski's discussion of power leads him finally to Weber's conclusion that power in modern times stems mainly from the ability to establish, and to enforce if necessary, certain special *rights* relative to others. As already noted, these rights are embedded in legal statutes for the most part, although the informal forces of custom, convention, and traditional beliefs or prejudices frequently supplement the official bases for power differences (Lenski, 1966: 32, 89). Once again, the parallels to Weber here are clear. For Lenski, the most prominent illustration of the legal basis for power is the formal right to private ownership of productive property. In addition, however, there are numerous examples from societies past and present of the legal establishment or denial of power for different races, sexes, religions, and so on. The supplementary and informal influence of nonlegal conventions and prejudices is also apparent when, for example, certain racial minorities or women are still excluded from positions of privilege, even though there are no legal or official barriers to block their access.

From the foregoing discussion, it should be apparent that Lenski shares with Weber a generally pluralist view of both power and inequality. Power derives from a combination of control over coercion and access to legally or conventionally sanctioned rights. Consequently, power differences can generate a wide range of social inequalities, depending on the variety of social categories that come to be accorded or refused such rights. Thus, a fundamental feature of Lenski's approach is that inequality is *multidimensional*, involving multiple criteria for ranking groups or individuals in terms of their power and, thus, their material privilege and prestige. The exact criteria for ranking obviously vary by country and historical period but usually include such diverse bases as property ownership, occupation, education, religion, ethnicity, race, gender, and age.

In many respects, Lenski's multidimensional view is both the key contribution and the one significant difficulty in his overall approach to social inequality. Predictably, Lenski sees Weber as the pioneer of this view, although Pitirim Sorokin also deserves mention (Lenski, 1966: 18; see Sorokin, 1927, 1947). However, subsequent critics argue that Lenski's multiple dimensions are different from Weber's delineation of class, status, and party in several key ways (e.g., Parkin, 1972; Giddens, 1973; Hunter, 1986). In particular, Weber's scheme is not explicitly intended to represent different ways for *ranking* people in society but rather to suggest the various bases around which interest groups may coalesce in society's continual power struggle (recall Chapter 3). Another way to express this is to note that Lenski is more clearly a *stratification* theorist than is Weber and that, compared with Weber, he may be more concerned with *distributive* inequalities in material and other rewards than with the *relations* between groups that underlie this distribution (1966: 2–3, 84–86).

Nevertheless, even if there is some confusion of these points in Lenski's analysis, the one undeniable and important similarity between his multiple dimensions and Weber's ideas of class, status, and party is that they all derive from the central concept of power. This makes it possible to conceive of both distributive and relational inequalities in terms of one key idea. It also suggests that our understanding of virtually all social hierarchies, whether they refer to class, race, gender, or any other social criterion, can be tied to the analysis of power, of access to enforceable rights. Whether these multiple forms of inequality are labelled "dimensions" or some other term is, in a sense, immaterial. One of the most insightful and useful aspects of Lenski's approach is precisely his recognition that such multiple forms exist and that, in one way or another, they are manifestations of power. This modified version of Weber's conception takes us significantly closer than earlier analyses to a *general* strategy for conceiving of social inequality.

Lenski on Class Structure

The one real difficulty with Lenski's multidimensional perspective is its treatment of the concept of class or class structure. At times, his discussion is not unlike Weber's once again, for Lenski visualizes a pluralist class structure, composed of a dominant propertied class, a subordinate working class, and a range of middle classes, as typical of capitalist countries like the United States. If there is a notable difference from Weber at all here, it is Lenski's tendency, as in Dahrendorf's early work, to be more convinced of the reduction of class inequality and the dispersion of wealth in modern society (1966: 308, 338–382).

Unfortunately, rather than confine his use of the term class to these standard economic concerns and treat class as one of the multiple dimensions of inequality, Lenski introduces some confusion by labelling all such dimensions as different types of class systems. In other words, he perceives "property classes," "educational classes," "ethnic classes," "religious classes," and so on, as constitutive of the overall class system (1966: 74–82). This strategy is again somewhat similar to Dahrendorf's view of class, since all power structures come to be defined as class structures, leading to a proliferation of different "classes" in society. Use of this terminology confounds class with power and risks making the class concept largely meaningless. The saving factor here is that we can expunge this weakness from Lenski's approach without abandoning its many useful elements.

The Role of the State

Given Lenski's focus on law and the formal institution of power, it is not surprising that, like Weber once again, he places considerable stress on the role of the state in modern systems of inequality. It is the state and its agencies that are responsible for creating, administering, and occasionally imposing by force the laws and formal rights that give power to some and not others. Because of this key role, and because of "the tremendous increase in the functions performed by the state," one might expect that the state's leaders would be central players in the overall power structure. In fact, Lenski does concur with Weber that power has been increasingly concentrated in the government, especially its growing bureaucracies; moreover, he also seems well aware that a dominant economic elite is typically able to turn the power of the state to its advantage on most key issues (1966: 304, 310, 342).

And yet, in the end, Lenski seems more optimistic than Weber that the cage of state bureaucratic control can be escaped and that exploitation by the ruling class can be reduced. To Lenski, the complexities of modern society, and the consequent intervention of the state into numerous spheres of life, have led to a greater, not a lesser, dispersion of power. The elaborate machinery of government has itself created diverse centres of jurisdiction and administration. Of course, the economic elite, as well as other dominant factions or interest groups in the ethnic, racial, gender, religious, and other hierarchies of society, may continue to gain disproportionate advantages from these agencies and the laws they uphold. But there is a limit beyond which such concentration of privilege cannot proceed without provoking public outcry, and it is the state structure itself that is now the key vehicle in this process. The people can oppose the policies of the ruling factions and resolve their own internal disputes through the state government, since it "has become the object of a never ending struggle between a variety of organized groups which, in their totality, represent the special interests of most of the population" (1966: 314–318). Here Lenski seems to share Dahrendorf's confidence that the power struggle can be legitimately institutionalized and acted out, not so much *by* the state as *through* the state. Here too, Lenski's perspective is somewhat similar to that of Dahrendorf, who sees government ideally as a "mere switchboard of authority" (Dahrendorf, 1959: 306). We shall encounter such images of the state again in subsequent sections of this chapter.

The Future of Social Inequality

As might be expected, Lenski's projections concerning the future suggest a guarded optimism, the same middle course that characterizes his overall analysis. Lenski devotes much of his discussion to an investigation of the historical trend of inequality. His central conclusion is that social inequality has tended to increase since primitive times but that there is reason to believe this trend is now reversing with the rise of advanced industrial societies. Among the many factors contributing to the alleged reversal are significant improvements in technology, which have made possible a vast increase in surplus wealth, and the spread of democratic ideas, which have captured the imagination of most people. These and other developments make it possible to redistribute both privilege and power to more and more citizens in the modern era (1966: 308–317, 428–430; see also Nolan and Lenski, 1996; Lenski, 1996, 2005).

Such positive expectations sound in some ways like Dahrendorf's favourable predictions for postcapitalist society. However, Lenski's views are tempered by his agreement with some of the assumptions that conservatives have espoused about humanity and social organization, which make the complete elimination of inequalities in power and privilege unlikely in *any* society. Lenski retains his conviction that social rankings are to some degree inevitable because of people's "natural tendency to maximize their personal resources" at the expense of others. In addition, hierarchies seem unavoidable given the importance of coordinated decisions in complex social structures and the differences in power or authority that result. Ultimately, then, the reduction of political and economic inequality should continue in the long term but "will stop substantially short of the egalitarian ideal in which power and privilege are shared equally by all members of society." In Lenski's opinion, radical thinkers who see socialism as the path to complete equality are destined to be disappointed (1966: 327, 345; 1980: 10; 2005: 135, 216–218).

Summary Observations

Our purpose in this portion of the chapter has been to review Lenski's perspective on social inequality as the relationship between power and privilege. Despite certain difficulties in his conception of class, a weakness that his perspective shares with Dahrendorf's approach, Lenski's attempted synthesis of radical and conservative views provides several crucial insights. In particular, his multidimensional strategy suggests an initial means for thinking about social inequality as a *generalized* phenomenon, a process that is critical for understanding class relations but that arises and endures in numerous other forms as well. Moreover, in stressing that the differential power to enforce rights is the common thread linking all these manifestations of inequality, Lenski achieves some significant advances over his contemporaries.

One of the best indicators of Lenski's sustained influence in the study of inequality is the continued application of his theoretical perspective in recent research. Most notable in this regard is the considerable evidence supporting Lenski's general proposition that future reductions in social inequality in most societies are closely linked to sustained industrial development and economic growth (e.g., Haas, 1993; Firebaugh and Beck, 1994; Nielsen, 1994; for some debate and discussion, see Patocka, 1998; Krymkowski, 2000; Blumberg, 2004; Lenski, 2005). Lenski's important contributions to the analysis of power and inequality should be kept in mind as we move through the remainder of this chapter.

NICOS POULANTZAS

In the more than one hundred years since Marx's death, there have been countless attempts by Marxist scholars to build on the ideas expressed in his extensive, but largely unfinished, writings. Debates continue to arise over which of these attempts represent the best or most promising efforts at reorienting or redirecting Marxian theory in recent years (e.g., Przeworski, 1985; Elster, 1985; Wood, 1986; for reviews, see Therborn, 1986; Levine and Lembcke, 1987; Burris, 1987; Mayer, 1994, Wright, 2005). However, while not all observers agree, it has been argued that the most significant of these recent discussions have primarily been those that connect in some way to Louis Althusser, a Marxist philosopher who has questioned the simplistic readings of Marx by certain early Marxists (e.g., Althusser, 1969, 1976; Althusser and Balibar, 1970). Within sociology, Nicos Poulantzas and Erik Olin Wright, whose works will be reviewed in the next two sections of this chapter, arguably have been the most successful in applying a broadly similar interpretation (see also Hindess and Hirst, 1975, 1977; Carchedi, 1977, 1987).

Although Poulantzas died in 1979, many contend that the power of his thought endures. His theoretical contributions are said to signify a "turn of the intellectual tide," a "paradigm shift" in the analysis of class structure (Therborn, 1986: 97). He has been described as "the single most influential Marxist political theorist of the post-war period" (Jessop, 1982: 153; see also Jessop, 1985: 6). Even during a period of history in which Marxist-inspired communist regimes have either fallen or gone into serious decline around the world, there are those who argue that Poulantzas's Marxist analyses of class structure and the capitalist state remain as significant and as original as ever (see Jessop, 1991).

Like Althusser, Poulantzas is notable for rejecting the dogmatic economic determinism of some other Marxists, those who completely reduce social analysis to an investigation of capitalist relations of production. While this issue is crucial, in Poulantzas's view there are other forces operating within society that must be grasped if the Marxist goal of revolution is to be realized. Moreover, Poulantzas argues, it is clear that "Marxism alone cannot explain everything," that the decisive role it plays in explaining social processes can still be enhanced by other valid insights (Poulantzas, 1978: 23).

With this in mind, Poulantzas seeks to supplement and redirect the traditional Marxist treatment of class, the capitalist state, and several other topics of importance for the study of social inequality. Despite certain obscurities and inconsistencies in Poulantzas's formulation, it is instructive to review and evaluate his attempts to accommodate the complications in modern systems of inequality with the essentials of Marx's original analysis. As with the other theorists we have considered, our review will focus on four principal concerns: the concept of class or class structure; the meaning and significance of power in social hierarchies; the role of the state, especially in capitalist society; and the prospects for inequality in the future, with particular reference to the possible transition from capitalism to socialism.

Classes in Contemporary Capitalism

Like many recent Marxists, Poulantzas conceives of class primarily as a set of "places" in a structure, although he sometimes uses class to refer to the "men" or "social agents" who fill these places (Poulantzas, 1973b: 27; Poulantzas, 1975: 14, 203). The distinction between classes as people and classes as structured locations is roughly paralleled in the structural-functionalist insistence that stratification refers mainly to the ranked status-roles which individuals occupy, rather than to the individuals themselves (Davis and Moore, 1945). While such a distinction is artificial, to Poulantzas and others it is seen as crucial for demonstrating that the *distribution* of people into classes is really a separate problem from the more pressing issue of how the structured *relations* between class locations are generated in the first place. For, even if we could imagine a capitalist society where people have an equal chance to move up or down in the class structure, this mobility would do nothing to alter the structure itself or the exploitative relations built into it (Poulantzas, 1975: 33).

In any case, both the class places of capitalism and the people within them are defined principally by their location in the productive, or *economic,* sphere. In addition, however, Poulantzas maintains along with

Nicos Poulantzas,
1936–1979

Reprinted by permission of Nicos
Poulantzas Institute.

Althusser that both Marx and his more discerning followers recognize how *political* and *ideological* forces can simultaneously act with the central economic factor in the formation of classes. The simultaneous nature of economic, political, and ideological influences on the class structure is important to stress here if we are to avoid fruitless "chicken-and-egg" debates over whether the political and ideological super-structures are "caused" by the economic infrastructure or vice versa (recall Chapters 2 and 3). To Poulantzas, it is important to retain Marx's key insight that economic forces are decisive "in the last analysis" for shaping classes and the other structural relations within capitalism. However, such a view is far different from the vulgar Marxist stance that the political and ideological systems operate in perfect coordination with the economic and are automatically and completely determined by it (1975: 14, 25; 1978: 26).

In delimiting the class structure, then, Poulantzas begins with the standard Marxist claim that there are two basic classes in advanced capitalism, the bourgeoisie and the proletariat, and that these are divided primarily by their economic relationship, by the fact that the bourgeoisie exploits the productive labour of the proletariat for profit. However, the class structure is also sustained over time by political and ideological processes. Like most Marxists, Poulantzas generally identifies the bour-geoisie's political control with the governing and coercive power of the state, and sees the bourgeoisie's ideological influence operating in the religious, educational, and other structures of society. As well, though, Poulantzas uses these two terms in less conventional ways: bourgeois political control is also embedded in the economic system, in the "politics of the workplace" created by capitalist rights of supervision and discipline over labour; bourgeois ideological control is likewise present on the job, in its monopoly of scientific ideas and technical knowledge, which deprives most workers of the means of "mental" production. Poulantzas perceives in Marx's writings this same wider sense of what political and ideological control mean and how these forms of bourgeois domination conjoin with economic relations to reinforce, perpetuate, or "reproduce" the essential class dichotomy (1975: 227–236; see Marx, 1867: 331, 361).

Having affirmed the Marxist view that the essential split between capitalists and workers is not blurred but rather highlighted by political and ideological forces, Poulantzas nevertheless notes that political and ideological factors can act to generate secondary splits *within* the two major classes, in the form of "fractions" or "strata" (1975: 23, 198). In the working class, for example, internal fragmentation can arise because certain workers have "political" control stemming from their supervisory or administrative duties, or because some occupations involve "ideo-logical" control based on their greater access to special information, knowledge, or educational skills (1975: 15, 245). Even in the bourgeoisie such political and ideological disunities can occur, because, despite their common interest in exploiting the working class, large-, medium-, and small-scale capitalists, as well as industrial, commercial, and finance capitalists, often break ranks due to their differences in political clout or policy preferences. Capitalists may also vary in their ideological commitments. For example, some truly believe in such ideas as fair competition and free enterprise, but others prefer a situation of complete monopoly control, along with the large and secure profits such monopolies can produce. These and other divisive tendencies often breed competitive or even antagonistic relations within capital, calling into question "the mythic image of the bourgeoisie as an integrated totality" (1975: 139; 1978: 143–144). This is not to deny the truth of Marx's two-class system, but to attack any simplistic representations of it.

Still, there is one aspect of Poulantzas's discussion of class fractions that has raised some questions about the dichotomous view of class and caused some controversy with other recent Marxists. Poulantzas identifies various class places near the boundary between bourgeoisie and proletariat. These class positions do not fall clearly into either class because they resemble the bourgeoisie on some political, ideological, and economic criteria but resemble the proletariat on others. In the end, he subsumes these intermediate locations under Marx's term, the *petty bourgeoisie,* to signify that they are, in a sense, linked to the bourgeoisie but play a petty or marginal role relative to this class. This designation does not completely solve the problem of classifying these positions, since they are themselves a heterogeneous mix, fragmented by political, ideological, and economic differentiation. Hence, out of this mélange Poulantzas identifies two major subcategories: the "old" or "traditional" petty bourgeoisie, a declining category of independent owners and craftspeople that both Marx and Weber refer to; and the "new" petty bourgeoisie, a growing array of technicians, supervisors, salaried "white-collar" employees, and "tertiary" wage earners (1975: 193, 208, 285–289).

It is Poulantzas's designation of a new petty bourgeoisie in advanced capitalism that has generated much of the debate over his work, both inside and outside Marxist circles. His new petty bourgeoisie represents a significant departure from the views of other Marxists, most of whom treat such positions as working class because their occupants are both propertyless and dependent on the capitalists for wages (e.g., Mills, 1951; Braverman, 1974; for discussion see Rinehart, 2006). Poulantzas concedes this dependence to some extent and notes that the new petty bourgeoisie, like the proletariat, is exploited by the capitalist class. Nevertheless, the supervisors, engineers, and other segments of the new petty bourgeoisie share considerable political and ideological ties with the capitalists, in Poulantzas's sense of these terms. That is, members of the new petty bourgeoisie tend to have some "political" control because they often direct the activities of workers in the productive setting, as well as "ideological" control because they typically possess technical knowledge and expertise denied to the proletariat. Their provisional resemblance to the bourgeoisie is further enhanced by their economic function, which produces no surplus value, in Poulantzas's view, and is paid for out of the surplus generated by the proletariat. In other words, they share with the capitalist class the bourgeois trait of being "unproductive" (1975: 210–216, 235–250).

Other Marxists remain unconvinced by Poulantzas's distinction between productive and unproductive activities, because so many class locations are mixtures of both and because it is not clear that technical knowledge and supervision are unproductive resources or attributes, in the sense of being irrelevant to the creation of surplus. Besides, all such activities are proletarian to the extent that they place their occupants in a situation of exploitation by capitalist employers. The most distressing aspect of Poulantzas's scheme, for some Marxist thinkers, is that his new petty bourgeoisie would form the largest single segment of the class structure, leaving a comparatively small and presumably less significant working class to fight for the overthrow of capitalism (e.g., Wright, 1978: 46–53; Wood, 1986: 42). Poulantzas himself sees in this no threat to revolutionary action, since the new petty bourgeoisie is likely to increase its proletarian allegiances and weaken its ties to capital with time (1978: 242–244).

Even so, Poulantzas's overall portrayal of the capitalist class structure raises difficulties for any Marxist who is otherwise sympathetic to his analysis. This is primarily because, in the end, Poulantzas comes surprisingly close to an essentially Weberian viewpoint. Like Weber, he wishes to acknowledge both the fundamental truths in Marx's two-class model and the intricacies that are overlooked if it is taken too literally. Like Weber, he is also faced with an infinite number of classes if he draws all the political, ideological, and economic distinctions that are possible within each class. Finally, like Weber, he ultimately settles on four key categories: the bourgeoisie, the proletariat, the traditional petty bourgeoisie of small owners, and the new petty bourgeoisie of salaried white-collar, technical, and supervisory personnel (recall Chapter 3). The major difference, of course, is that Poulantzas rejects Weber's treatment of these latter two categories as distinct *middle classes* in their own right, for this would contradict the basic Marxist precept that there are only two principal classes in advanced capitalism (1975: 196–199, 297).

Poulantzas on Power

It has been typical of Marxist discussions to avoid any detailed attempt to conceptualize the notion of power, perhaps because Marx himself devoted little effort to this task. Poulantzas does use the idea of power in his analysis, but in an ambivalent manner. His initial definition treats power solely as the capacity of a *class* "to realize its specific interests in a relation of opposition" to another class (1978: 36; see also 1973a: 99; 1975: 277). Poulantzas seems to be affirming the conventional Marxist preoccupation with class issues here, since by this definition all situations of opposed interests that do not involve class are not worthy of being called power struggles. This usage poses conceptual problems that are almost the reverse of those found earlier in the approaches proposed by Dahrendorf and Lenski. As we have seen, the latter two writers label all power relations as different types of class relations, thereby making class a meaningless idea. Poulantzas chooses to ignore all power relations that are not class relations and thus makes power the superfluous term.

Yet, on closer inspection, the ambivalence and inconsistency in Poulantzas's view of power are evident, for elsewhere he reveals a wider sense of power that is intriguingly similar to Weber's conception. First of all, his definition of power as a capacity to realize interests despite opposition is very much like Weber's view, apart from its restriction to class issues (see Weber, 1922: 53). Moreover, even this restriction appears to disappear at times, since Poulantzas allows that "power relations stretch beyond class relations" to include problems of race, gender, and so on (1978: 43–44; see also 1975: 305–306). On these grounds, at least, Poulantzas's neo-Marxist approach to power, like his analysis of class structure, seems compatible with a limited form of Weberian pluralism. If there is a real quarrel, it is with more extreme pluralists who fail to recognize that economic power is central "in the last instance," and that "class power is the cornerstone of power" in the other areas (1973a: 113; 1978: 44).

The Capitalist State

Poulantzas distinguishes himself from many early Marxists by his attempt to include the state as an important factor in the structure of advanced capitalism. Unfortunately, his discussion is often obscure, primarily because he once again walks a nebulous line between orthodox Marxism and some variant of Weberian pluralism. To begin with, his definition of the state as the "condensation" or "fusion" of class struggle is too cryptic to be very informative. He seems to mean that the state is an organizational shell for society, a vast framework of rules and principles that ensure bourgeois domination in all those spheres not immediately part of the economy (1973a: 53–55; 1978: 26–30; see also Jessop, 1985: 54). At a concrete level, the state thus consists of a network of organizations or "apparatuses" that are of two related types: the political apparatuses, including the executive, legislative, judicial, civil-service, police, and military arms of government; and a range of ideological apparatuses such as education, the mass media, and so on. This portrayal resembles the conventional views of Weber and even Durkheim, except for Poulantzas's curious inclusion of religion, the family, and other structures within the state's ideological apparatuses (1975: 24-25). As others have noted, this means that virtually everything but the economy, or the material production process, is subsumed under the capitalist state according to Poulantzas (Giddens and Held, 1982: 193).

Given this image of the state as an all-encompassing framework for bourgeois domination, one might suspect that Poulantzas accepts the crude economic determinism of those Marxists who see the state as an appendage or tool of the capitalist class. However, Poulantzas wishes to dissociate himself from this viewpoint while at the same time avoiding charges of pluralism. His compromise is to argue that the state apparatuses are not totally independent operators in the capitalist power structure but nevertheless are *relatively* autonomous from the bourgeoisie (1973a: 256; 1975: 158; 1978: 13). The heads of the state apparatuses are themselves bourgeois for the most part, while the rest of the state personnel are mainly members of the new petty bourgeoisie, white-collar civil servants with varying political and ideological ties to the capitalist class (1975: 187; 1978: 154-155).

Such inbred allegiances mean that the state will uphold the general interests of the bourgeoisie. Nevertheless, the state is too large and, like the class structure, too fragmented by special interests to exercise a unified political will on all issues. On the contrary, the complex bureaucratic amalgam of state agencies generates an intricate mix of "diversified micropolicies," many of which are "mutually contradictory" (1978: 132-135, 194). The overall result is that certain specific policies of the state may produce "short-term material sacrifices" by the bourgeoisie: being required to pay for the improved health and safety conditions of workers; contributing through taxation to public education, social security, or unemployment benefits; and so on (see Jessop, 1985: 55). Still, these actions by the state really benefit the capitalists in the end, by defusing potential proletarian revolts, promoting a more compliant and dependable work force, and thus securing the "long-term domination" of the bourgeoisie (1978: 30-31, 184-185; see also Jessop, 1985: 61).

Poulantzas's determination to find a middle ground, between portraying the state as a tool of capital and depicting the state as an independent force in society, ultimately leads to his paradoxical claim that the state, on the one hand, is the very "center" of power in capitalism but, on the other hand, "does not possess any power of its own" (1973a: 115; 1975: 81; 1978: 148). Given his peculiar conception of power as an exclusive aspect of class relations, it does of course follow that the state by definition has no power, since it is not a class (see Connell, 1979). And yet Poulantzas must reconcile this idea with his view that the state is now the central setting in which power is exercised, because of massive state involvement in social services, public administration, and, increasingly, the operation of the capitalist economy itself (1975: 81; 1978: 168). His insistence that the state has no power is justified primarily by the fact that the capitalist state does not control the means of production, even if it does have an expanded role in taxing and spending the surplus generated by the economy. Besides, state incursions into the economy never go beyond certain limits, for fear of eroding the profit motivation of the bourgeoisie on which the state's own operating funds depend. Often, in fact, state involvement is really a desperate attempt to help the bourgeoisie out of economic crisis or depression. It represents the dilemma that state leaders face in trying simultaneously to placate a discontented working class and a capitalist class concerned mainly with its own gains. In short, the state is far from omnipotent under advanced capitalism but instead stands with "its back to the wall and its front poised before a ditch" (1978: 191–192, 244).

At least one key difficulty with this assessment of state power should be noted. Poulantzas's own definition of power indicates that it involves the realization of a faction's interests in the face of opposition. But control of material production does not exhaust the means by which interests can be realized or opposition quelled. In particular, we should not overlook the use of coercive force or repression to exact compliance from any opponents. In fact, Poulantzas follows Weber, Lenski, and others in acknowledging the fundamental role of repression in *all* power relations and notes that power in its most basic form entails physical force, quite literally "the coercion of bodies and the threat of violence or death" (1978: 28–29; see also 1973a: 225). But who has the capacity to employ repressive power? Clearly, it is the state that formally controls this resource, especially through the legalized actions of the police, the military, and the official justice system. In addition, the activities of virtually the entire state bureaucracy enjoy the legal sanction and, if necessary, the coercive aid of these agencies. Poulantzas seems to recognize this point and credits Weber with being the first to establish it (1978: 80–81). However, Poulantzas fails to appreciate fully that this makes state power a distinct force to be reckoned with in society.

Of course, in rejecting Poulantzas's claim that the state has no power, we should not simply take the opposite stance that the state in capitalism has a total monopoly of power. Thus, Poulantzas is correct to note, along with Lenski and others, that organized state repression is subject to limitations, particularly in nominally democratic societies. The laws that establish state powers in modern times can also restrict them, delimiting the state's jurisdictions and spelling out popular rights. Blatant disregard for these limits to power risks public outrage and possible open rebellion against state control (1978: 31–33, 82–83). The crucial point to stress is that the state's powers are indeed constrained but that neither these restrictions, nor the state's ceding of ultimate economic control to the capitalist class, leaves the state apparatuses without inherent influence. Perhaps the key weakness in Poulantzas's entire perspective is that he emphasizes the relative autonomy of the state and yet cannot admit the distinct state power that this argument implies. In addition, his interest in the abstract framework of state apparatuses leads to an insufficient acknowledgment that these apparatuses are also concrete organizations, run by real people who have considerable say in the running of society.

The Prospects for Socialism

As a committed Marxist, Poulantzas wishes primarily to understand and promote the transformation of capitalism into socialism. Here Poulantzas addresses a question faced by all modern Marxists: why have so many attempts to create an egalitarian society through socialist revolution led to systems bearing little or no resemblance to Marx's version of communism? Interestingly, Poulantzas believes the problem lies with the state (see Jessop, 1985: 118–119). Such perversions of socialist principles as Stalin's Russia, for example, occurred because of the misguided belief of Lenin and others that an utter smashing of the capitalist state can solve the administrative problems of socialist society. Instead, the resulting administrative void is filled by a "parallel" socialist state, one that does not wither away but too often is more resilient, more bureaucratic, and more repressive than its bourgeois predecessor (Poulantzas, 1978: 252–255).

Poulantzas's alternative strategy is to work within the existing capitalist state to transform it gradually. Briefly, this strategy entails the spread of trade-union activities, workers' political parties, and other new forms of "direct, rank-and-file democracy" (1978: 261). In addition to the working class, a key force for change is the new petty bourgeoisie, especially many middle- and lower-level state employees, whom Poulantzas expects to ally with the proletariat as economic crises occur and their material conditions worsen (1978: 242–244). To be sure, the eventual use of force cannot be ruled out as potentially necessary to final success. Moreover, it remains to be seen whether current leaders on the left are yet capable of enlisting and organizing mass action. But, in any event, no truly democratic socialism is possible without this base of "broad popular alliances" to counteract totalitarian tendencies (1978: 263).

Even in his prescription for the future, then, Poulantzas shares a certain similarity with Weber, for both theorists have a distrust of socialism that is really bureaucratic *statism* in disguise, and both advocate working within existing bourgeois structures to change them. Of course, Weber and Poulantzas are unalterably opposed on both the extent and the vehicle of social change. Whereas Weber supports rather minor revisions to bourgeois liberal democracy and sees an enlightened political leadership as the key force for curbing bureaucratic domination, Poulantzas desires, like Marx, the fundamental shift from bourgeois democracy to democratic socialism and considers the mass of the people as the principal factor in this transformation.

Summary Observations

This section of the chapter has addressed Poulantzas's neo-Marxist perspective. We have seen that Poulantzas's image of inequality, as reflected in his analyses of class, power, and the capitalist state, falls somewhere between conventional Marxism and a limited form of pluralism broadly suggestive of Weber. In this way, Poulantzas is representative of a large number of modern Marxists who have become engaged in "a prolonged dialogue with the ghost of Weber" (Burris, 1987: 67). Although some Marxists worry that this dialogue has resulted in a "Weberianization of Marxist class analysis" (Levine and Lembcke, 1987: 6), there are some advantages to building linkages between the two formulations. In Poulantzas's case, the outcome is a perspective that sustains Marx's basic emphasis on control of economic or material production as the key source of power and inequality in society, but that also acknowledges the important political and ideological forces that sometimes escape the notice of other Marxists. Where there are weaknesses in Poulantzas, they arise mainly from his lack of conceptual clarity in some areas and his rather ambivalent handling of the problem of state power. We are left with a contradictory impression of a state that pervades more and more of our lives but that somehow has no substantive power of its own. The major omission here is the realization that the state personnel are themselves social actors with powers of legal and physical coercion at their disposal.

ERIK OLIN WRIGHT

The second major neo-Marxist scholar to be considered in this chapter is Erik Olin Wright. Like Poulantzas, and Marx himself, Wright's primary concern has not been to devise a comprehensive theory of social inequality, but rather to investigate the prospects for socialist revolution, given the inherent characteristics of advanced capitalist societies (Wright, 1978: 26). In doing so, Wright has devoted much of his research to explaining the nature of capitalist class structures and class relations (see, e.g., Wright, 2005). Some observers suggest, in fact, that his work represents the most significant and sweeping endeavour of all current attempts to formulate social class and to make the concept amenable to empirical analysis (e.g., Therborn, 1986: 99). At the same time, however, Wright's class analyses have also generated some new and important insights into how social inequality more generally can be conceived and understood.

As we shall see, Wright's conceptualization of class structure in contemporary capitalism has undergone some notable changes since his original formulation (Wright, 1976, 1978, 1979). In his most recent writings, Wright has also distinguished himself more clearly from the "structuralist" Marxism of writers like Althusser and Poulantzas. He sees his own work as part of a newer "analytical" Marxism that seeks to reconstruct contemporary Marxist thought in several important ways (e.g., Wright et al., 1992; Wright, 1994, 1997, 2005; see also Mayer, 1994).

Even so, it is clear that Wright shares some similarities with both Poulantzas and Althusser. Most notable perhaps is the common goal of developing a revitalized and sophisticated reorientation of Marxist theory. A recurring theme in Wright's work, especially his latest analyses, is the rejection of the many dogmatic or doctrinaire aspects of some traditional Marxism (e.g., Wright et al., 1992: 181; 2005: 4, 15). In Wright's estimation, truly effective Marxist scholars must be "open to continual reassessment of their own theoretical positions, acknowledging their theoretical failures as well as arguing for their successes" (Wright, 1994: 184). At the same time, he also is one of the leading proponents of the need to make Marxist theory applicable to concrete research questions, even if the conceptual issues raised in theoretical debates have yet to be completely resolved. Wright takes the practical view that there is much important research to be done, and "it is better to forge ahead ... with somewhat less certain concepts than to devote ... an inordinate amount of time attempting to reconstruct the concepts themselves" (Wright, 1997: xxix).

Wright devotes his attention to several of the same issues that are of crucial concern to Poulantzas and that tie closely to the analysis of inequality in advanced societies. For our purposes, the topics of note are the capitalist class structure, especially those anomalous "places" or

**Erik Olin Wright,
born 1947**

Reprinted by permission of Erik Olin Wright.

"locations" that are somehow distinct from the bourgeoisie and proletariat; the origins of power or domination; the role of the capitalist state;
and the outlook for a socialist transformation of capitalism in the light of
these other considerations. We shall first review his original conception of
class and inequality in detail and then examine some of the key changes
he has introduced in his most recent writings.

The Capitalist Class Structure

Like Poulantzas, Wright wishes to retain the essential two-class model of
nineteenth-century Marxism, but in an updated form that incorporates
important modifications and extensions of recent times (see, e.g., Wright
and Perrone, 1977). However, while Poulantzas sees these developments
in terms of "fractions" within the two main classes, Wright uses a more
systematic strategy that remains consistent with the classical Marxian
concern with ownership and control of the capitalist economy.

According to Wright, capitalism in its most "abstract" or "pure"
sense does indeed generate only two classes: those in control of economic
production, the bourgeoisie, and those excluded from control of production, the proletariat. However, in real capitalist societies, complications arise, first, because a third class, the old petty bourgeoisie, continues
to exist (albeit in a declining form) and, second, because command over
the economy is a more complicated issue than it was in the previous
stages of capitalism (Wright, 1985: 7–8; see also Wright, 2000: 961).
Wright's earliest attempt to conceptualize classes is based on the premise
that control of economic production in modern capitalism is divisible into
three key elements: (1) "real economic ownership," which is most crucial
and refers to control over all the economic surplus—the profits, products,
and other resources of capitalism; (2) command of the physical apparatus
of production, which entails supervisory control over the machines,
factories, and so on, that are used to make products; and (3) command of
labour power, which refers primarily to supervisory control over workers
(1978: 73; 1979: 24).

Using these three criteria, Wright proposes that the present-day
bourgeoisie be defined as the set of positions or class locations in which
all three types of control are retained, while the proletariat includes those
class locations where all three types of control are absent. The third class
in capitalism, the petty bourgeoisie, is a carry-over from earlier times and
includes positions filled by people who, as the owners of small businesses,
for example, generate and control their own surplus (control 1) and manage
their own enterprise (control 2) but do not employ workers (control 3).

As for the rest of the class structure, it is composed of positions that,
strictly speaking, do not form classes at all because they have some
elements of economic control but not others. Wright sees these as
"contradictory" locations arrayed between the three main class clusters,

Figure 6.2
Wright's Original Model of Capitalist Class Relations

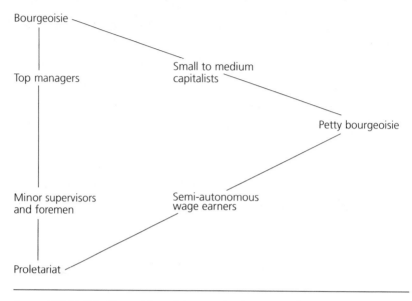

Bourgeoisie

Top managers

Small to medium
capitalists

Petty bourgeoisie

Minor supervisors
and foremen

Semi-autonomous
wage earners

Proletariat

Source: Wright, E.O. *Class, crisis, and the state.* London: New Left Books/Verso U.K. p. 84. Reprinted by permission of Erik Olin Wright.

as in Figure 6.2. Thus, between the bourgeoisie and proletariat are all those positions filled by people who resemble proletarians, in that they do not own or control the economic surplus (control 1) but have varying degrees of supervisory command over both the physical plant and the employees of the enterprise (controls 2 and 3). Along this line, top-level managers are closest to the bourgeoisie, since they control the most workers and large segments of the apparatus, while toward the proletarian end of this continuum are minor supervisors and foremen, who oversee only small sectors of production and a few workers.

The other two ranges of contradictory class locations include small-to medium-scale capitalists, positioned outside the bourgeoisie and toward the petty bourgeoisie primarily because they employ relatively few workers; and "semi-autonomous" wage earners, a mix of scientists, professors, and other salaried professional or technical personnel who are located between the petty bourgeoisie and the proletariat because, like the latter, they employ no workers but, like the former, they have some command over the products of their labour and over the physical means used in their creation (controls 1 and 2) (1978: 80–81; 1979: 46–47).

In this innovative manner, Wright poses an alternative strategy for conceiving of places in the class structure that do not fit precisely into the

Marxist two-class model. Wright avoids lumping such positions into a single new middle class, as Dahrendorf does, or a single new petty bourgeoisie, as Poulantzas does. Instead, Wright locates these positions systematically along three separate ranges that vary according to both the type and the degree of economic control enjoyed by their incumbents.

And what of people employed in spheres outside the production process, in the political and ideological structures of society? Here Wright also shuns Poulantzas's simple treatment of all such positions as unproductive elements of the bourgeoisie or petty bourgeoisie. In classifying these locations, Wright's original formulation relies on much the same rationale as that used for the sphere of economic production. That is, within the political and ideological apparatuses of capitalism, class location can also be defined by control of production, but here production refers to political and ideological creations, especially the policies, laws, ideas, and beliefs generated and communicated by such structures. Hence, most of the political and ideological personnel are proletarians, since they are excluded from control over the creation and implementation of policies and ideas. A few at the top—political leaders, supreme magistrates, the heads of education, religion, and so on—control both creation and implementation and therefore can be thought of as members of the bourgeoisie. The remainder have some partial control over such matters and so occupy contradictory locations between the bourgeoisie and the proletariat. The one apparent difference between these class alignments and those within the economy itself is the absence of any political or ideological counterparts to the petty bourgeoisie (1978: 94–97; 1979: 54).

Apart from this discrepancy, though, Wright's first attempt to develop a new Marxist class categorization allows us to envision an array of political and ideological apparatuses, running side by side with the economic structure and representing all the major spheres of capitalist society. As depicted in Figure 6.3 (page 150), it is possible to conceive of the bourgeoisie as the intersection of positions at the top of all these structures, while the proletariat occupies the intersection along the bottom, and the various contradictory locations lie between these extremes (see also Wright, 1980: 190). It is noteworthy that this depiction bears some resemblance to conventional sociological portrayals of society as a series of interconnected structures, each engaged in a particular task within the larger system. This view enables us to conceive of class structure as a phenomenon that cuts across every sphere of social organization and that, in a sense, joins them all together.

Wright's Subsequent Reformulation of Classes
The conception of classes in contemporary capitalism that Wright first proposed has had a major influence on the thinking of both Marxist and non-Marxist scholars. Apart from the theoretical debate and discussion it

Figure 6.3
The Intersection of Class Relations across Economic, Political, and Ideological Structures

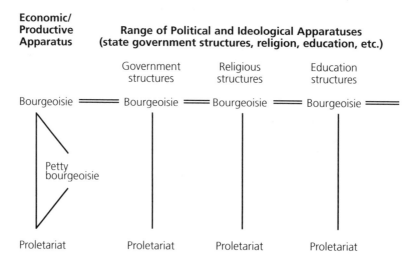

Economic/
Productive
Apparatus

Range of Political and Ideological Apparatuses
(state government structures, religion, education, etc.)

has stimulated, Wright's scheme has been widely adopted by other writers as perhaps the best available means for measuring or operationalizing class in empirical research (see, e.g., Grabb, 1980; Johnston and Ornstein, 1982; Hagan and Albonetti, 1982; Hagan and Parker, 1985; Baer, Grabb, and Johnston, 1987, 1990; Li, 1992; Veenstra, 2005).

Nevertheless, more recently Wright has proposed an alternative conception of classes to replace his original version. Wright's reformulation is partly a response to those critics who have argued that his first conception deviates in some key respects from the classical Marxian perspective. Wright agrees with such criticisms to some extent, including the claim that his initial approach departs from the exploitation-based definition of classes usually identified with Marx. Wright concedes that his first formulation, like most neo-Marxist classifications of the period, put undue emphasis on power or domination as the defining criterion for class differences (1985: 15, 56). For example, his use of various types of control over (or domination of) the productive process to define classes leads to a scheme that, in many ways, seems closer to a Weberian than to a Marxian conception of class.

In place of the old formulation, then, Wright has sought to devise a classification based more clearly on the concept of exploitation, whereby class distinctions arise from the "economically oppressive appropriation of the fruits of the labor of one class by another" (1985: 77; see also 1994: 43–46, 239). Building in part on earlier writings by John Roemer (1982),

Wright concludes that, in modern capitalism, the exploitation of one class by another emerges from three key processes or sources. In addition to the familiar appropriation of resources or assets by owners within the means of production, exploitation also occurs through the differential benefits that some people gain by having higher levels of skill or credentials ("skill/credential assets"), or by being more strategically placed in the organizational hierarchy of the productive process ("organization assets") (1985: 88). The various combinations of these three different forms of exploitation give rise to a twelve-category typology of class locations in contemporary capitalist society (see Figure 6.4, page 152).

Ironically, in spite of Wright's desire to develop a scheme that is more faithful to Marx's conception of classes, many contemporary observers, particularly other Marxists, have concluded that Wright's second typology is actually less consistent with classical Marxist thought than was his original approach (e.g., Meiksins, 1987: 51–57; Carchedi, 1987: 112–131; Kamolnick, 1988). First of all, rather than produce a set of categories that more closely resembles the simple dichotomy of bourgeoisie and proletariat reminiscent of early Marxism, Wright's reformulation actually generates a more extensive list of class categories than before, one that is more pluralist and, in effect, more Weberian than Wright's original approach. Wright appears to acknowledge this point to some degree. He notes, for example, that his scheme represents a "hybrid" of Marxist and Weberian class analysis and that, when doing empirical research, the categories he uses "can be interpreted in a Weberian ... manner" (Wright, 1997: 29, 37). Ultimately, however, Wright contends that the Weberian strategy of class analysis is subsumed under, or "nested within" the Marxist model, rather than the other way around (Wright, 2005: 27; see also Wright, 2002).

Some critics also suggest that the theoretical gains achieved by Wright's more recent classification seem limited at best, particularly in light of its greater complexity. Essentially, we still have the three main classes that form the triangular pattern in Wright's original scheme: bourgeoisie, proletariat, and petty bourgeoisie (Figure 6.2, page 148). The key difference is that, in place of the three ranges of "contradictory locations" linking these three classes, there now are nine new categories, all differentiated by whether or not the incumbents in these locations have a great deal, some, or none of the three types of assets (ownership, skills/credentials, and organization).

A second major criticism of Wright's newer classification concerns certain problems with his definition of exploitation. Take, for example, his view of skill/credential assets. According to Wright, skilled workers or those with special educational credentials actually exploit less skilled workers, at least indirectly. To Wright, those with skills or credentials can use these resources to earn higher incomes than other workers. The extra

Figure 6.4
Typology of Class Locations in Capitalist Society

Assets in the means of production

	Owners of means of production	Nonowners (wage labourers)		
				Organization assets
			+	
Owns sufficient capital to hire workers and not work	1 Bourgeoisie	4 Expert managers	7 Semi-credentialled managers	10 Uncredentialled managers
Owns sufficient capital to hire workers but must work	2 Small employers	5 Expert supervisors	8 Semi-credentialled supervisors	11 Uncredentialled supervisors	> 0
Owns sufficient capital to work for self but not to hire workers	3 Petty bourgeoisie	6 Expert nonmanagers	9 Semi-credentialled managers	12 Proletarians	–
		+	> 0	–

Skill/credential assets

money they receive amounts to an additional "rent" that skilled workers are able to extract from the productive process, over and above the surplus they generate themselves (1994: 252). In essence, then, this rent is acquired at the expense of less skilled workers, in somewhat the same way that the profits obtained from capitalist ownership are appropriated from the fruits of other people's labour. For this reason, Wright suggests that skilled workers occupy a mixed class location, being in a position where they are exploited by capitalist owners, but where they also exploit the labour power of other workers (see, e.g., Wright, 1994: 45–48).

To many critics, this way of thinking about exploitation is one that Marx himself would not accept. In particular, Wright's conception seems more distributive than the relational view of exploitation found in Marx's work. That is, skilled workers are judged by Wright to be beneficiaries of exploitation, even if they have no direct relationship with or control over the activities of other workers.

Interestingly, Wright agrees with this criticism to some extent, although at the same time he maintains that his newer typology is still an improvement on the previous version (Wright, 1989a: 10; 1994: 66; for debate see Wright, 1989b). Even so, Wright concedes that the practical advantages of the second conception over the original are questionable. He notes, for example, that the working-class categories in the two schemes "overlap so much that an empirical comparison is difficult to pursue" and that the recent version does not represent "a decisive empirical break with the previous approach" (1985: 187; see also Wright and Martin, 1987; Wright, 1997: 17–26). In other words, regardless of which of the two classifications is employed in empirical analyses, the results are very much the same. Some recent research, while critical of both approaches, suggests that Wright's original domination-based set of classes may still be empirically and conceptually superior to the later exploitation-based version (Halaby and Weakliem, 1993: 59; for debate see also Wright, 1993; Halaby, 1993).

It is a sign of Wright's flexibility of thought that, although he has relied on the exploitation-based scheme in his most recent empirical work (e.g., Western and Wright, 1994), he remains open to the prospect of revamping or modifying his approach yet again, if the advantages of such changes can be demonstrated (see Wright, 1994: 184). Increasingly, in fact, Wright now suggests that elements of both the old and the new approaches should be incorporated. He contends that domination, which is central to his original typology, and exploitation, which is the basis for his later classification, should both be seen as essential ingredients in any complete definition of classes. In his view, "exploitation without domination, or domination without exploitation, do not constitute class relations" (Wright, 1994: 60–61, 20, 63; 1985: 300; 2005: 25; see also Agger, 1992).

Power and Domination

Wright's discussion of the concepts of power and domination is less explicit and less detailed than his analysis of class. Still, as we have seen, power and domination clearly play a notable role in Wright's work, both in his original delineation of classes in capitalist society and in his current thinking on how best to define classes.

In his earliest writings, there is evidence that Wright's ideas about power largely conformed to those of most other Marxist writers. That is, although his work would occasionally imply that power is a phenomenon that operates beyond or outside of class issues (e.g., Wright, 1979: 197, 227–228), Wright tended to treat power and class as inseparable ideas for the most part. In this sense, Wright's original approach paralleled that of Poulantzas, who, it will be recalled, saw power as a capacity that, strictly speaking, only classes can possess or not possess. It will also be remembered, though, that such a view leads to serious conceptual difficulties for Poulantzas, since it means that nonclass forms of inequality (e.g., racial or gender inequality) are not understood as problems of power in their own right, but instead are treated as simply secondary consequences of power differences between classes.

Perhaps as a result of such difficulties in the standard Marxist approach to power, Wright's more recent discussions suggest a significant shift in his thought. Wright's treatment of power is now generally consistent with more conventional Weberian conceptions. In fact, Wright draws on the works of several Weberian or non-Marxist writers, including Lenski and Giddens, when developing his definition (e.g., Wright et al., 1992: 68–73; Wright, 1994: 23–24, 93). Wright currently defines power or domination as a differential capacity to achieve goals or outcomes, arising out of "a relational process of competition and conflict between contending individuals, groups, nations, etc." (1994: 24; see also Wright, 2000: 962). Such a definition makes it apparent that, in Wright's view, power is not a capacity that inheres only in class relations, but instead can involve various groups or collectivities differentiated along lines of gender, race, nationality, and so on. In contrast to some other Marxists, then, Wright recognizes the significance and distinctiveness of several "non-class forms of domination" (Wright et al., 1992: 75; Wright, 1994: 247–248; see also 2005: 22–23).

Of course, in spite of this shift in his approach to power and domination, Wright has not abandoned his essentially Marxist perspective on social inequality. Thus, he opposes those who believe that the acknowledgment of these other kinds of power relations must mean the "displacement of class from the centre stage" (1985: 57). He contends that class power still remains crucial for understanding contemporary social inequality, and that class is not just "one of many oppressions, with no particular centrality to social and historical analysis" (1985: 57; 1989a: 7).

Wright has suggested several reasons why class remains so central. Included among these is the classical Marxian idea that class relations and class struggle have a special "dynamic" quality that promotes development and social change in society, and that does so in ways that gender, race, and other social relations do not (see Wright et al., 1992: 74–76; Wright, 1994: 194–95). Although this claim may be debatable to many observers, the economic argument for the importance of class may be more compelling. Both Marxist and non-Marxist theorists seem to agree that, in capitalist societies at least, class structures and class relations are principal determinants in the production and accumulation of economic surplus. In turn, access to this economic surplus is usually pivotal, because whoever controls it can place clear limits on the range of possibilities for social change in other areas. Thus, according to Wright, the material resources that are generated out of the capitalist economy and its attendant class system are crucial, both as means and as ends, in the numerous "ethnic, religious, nationalist, and other non-class movements" that have arisen in modern times (1994: 65, 247–248; see also 2005: 24–25). In this sense, Wright believes, class always plays a "central, if not necessarily all-encompassing, explanatory role" in shaping other forms of inequality in contemporary societies (1994: 92, 210, 248; 1997: 1–3).

Even here, though, the more flexible and qualified nature of Wright's newer "analytical" Marxism leads him to the conclusion that, while class and economic issues remain central and pervasive, this is not to say that they must be the most important or significant factors in explaining all types of inequality or group conflict: "there almost certainly are situations in which gender or racial conditions more deeply stamp the subjectivity of actors and their conditions for struggle than class does" (Wright et al., 1992: 75). In the same passage, Wright goes on to say that, "while struggle for control of material resources" is one essential factor in understanding social inequality and social change, other influences, most notably "struggles over ideology and control of political institutions," may be equally or more significant in certain instances (see also Wright, 1997: 303–304, 544–545; 2005: 22–23).

Wright's recognition of these distinct economic, political, and ideological forms of struggle or control in society suggests considerable pluralism in his present approach to thinking about power and social inequality (see also Wright, 1994: 115–118). This pluralism is evident, as well, in Wright's inclusion of gender, race, and other nonclass bases for domination as examples of power relations. As we have seen, occasional hints of a similar tendency can be found in Poulantzas's thought. Nevertheless, compared with Poulantzas and most other contemporary Marxists, Wright's recent work represents a far more explicit rejection of the old economic determinism or "class reductionism" of some early

Marxism, the idea that economic control and class relations somehow explain everything. Over time, Wright has continued to develop and refine his theoretical position in this way. Although he retains an essentially Marxist perspective, at the same time, he consistently and increasingly has tempered it with a sophisticated and self-critical "intellectual pluralism" (Wright, 1994: 8; also Wright et al., 1992: x).

The Capitalist State

A similar tone and approach is evident in Wright's discussions of the capitalist state. Perhaps the first point to note is that Wright holds a generally conventional conception of the state. That is, in contrast to Poulantzas, he does not see the state as encompassing virtually everything but the economic structure of capitalism, but instead defines the state as mainly the government-run apparatuses of society: the political structure, its civil-service bureaucracies, and certain ideological arms such as public education and state-administered mass media organizations (Wright, 1978). Like Poulantzas, however, Wright sees the state structure as one that is constrained in its actions by the power of the capitalist class. The key reason is that the state generates little or no surplus of its own, and so relies primarily on the taxation of privately accumulated profits and wages for its funds (see Wright, 1994: 97; see also Offe and Ronge, 1975).

This is not to say that the state apparatuses are therefore unswerving servants of their capitalist benefactors. Again, Wright parallels Poulantzas in arguing that the state normally has a degree of autonomy from the capitalist class when it comes to setting and implementing government policies (Wright, 1994: 148–149). Like Poulantzas, he also notes the special "repressive capacity of the state" that derives from its formal control of the police and military (Wright et al., 1992: 45).

For these reasons, there is no doubt that sectors of the state can operate with some independence from the capitalist class and can also act against the wishes of the bourgeoisie on specific issues. Illustrations might include establishing occupational safety standards to protect workers from job hazards created by their employers, or fining companies found guilty of environmental pollution. Such behaviour by the state may arise from a genuine commitment to the general welfare, as well as from the need to appease discontent within the electorate or various interest groups in the public at large. At times, the state leadership may also try to protect the bourgeoisie from its own self-destructive internal conflicts, by negotiating or imposing settlements in cases of disputes among individual capitalists (e.g., Wright, 1994: 95).

These activities suggest a capitalist state with real power of its own. Even so, Wright's view is that there are also clear limits on the exercise of state power. The most important constraint is the vested interest that the capitalist state has in preserving "the goose that lays the golden egg"

(Wright, 1994: 101). This means maintaining a largely contented business community, possessed of the economic incentive to amass profits, employ workers, and thus ensure the state's vital operating revenues. A related point is that many government leaders have bourgeois connections themselves, and most state officials embrace a generally pro-capitalist ideology (see Wright, 1994: 151).

In seeking to promote a stable and expanding economy, however, the state in advanced capitalism is confronted with some serious dilemmas. In particular, Wright argues that the state leadership may be obliged to intervene in the economic sphere, often against its own will. Pressures for state intervention are said to occur, not because state leaders want to control the business sector, but because certain tendencies toward crisis and depression regularly disrupt the capitalist economic system. Such crises may be even more serious in advanced capitalism where, as Marx foresaw, economic control has become concentrated in a relatively few giant corporations. Whereas in the past economic depression usually led to the failure or absorption of small enterprises, now it is often large-scale businesses that are threatened in economic declines. Thus, when Chrysler in the United States or Dome Petroleum in Canada faced bankruptcy in the 1980s, major elements of the capitalist class were endangered, as well as the jobs of thousands of workers. In such cases, the state leadership is under pressure to step in and save the companies in question, so as to protect capitalist interests, avert popular unrest, and preserve the government's tax base.

But, in Wright's view, the state's actions in these situations usually involve short-term solutions that can create new problems. Bailing out weak companies, giving preferential treatment to troubled businesses, and increasing taxes to provide more unemployment benefits may simply put more burdens on the system. That is, such actions can involve subsidizing inefficient, unproductive elements of the economy at the expense of the healthy sectors. These policies can also drain surplus wealth and resources through the creation of costly state bureaucracies to administer the programs. Wright argues that these added burdens tend eventually to engender new crises, forcing the state to intercede still further in a continuing cycle of intervention that satisfies neither the bourgeoisie nor the majority of workers (1978: 156–163, 174–179).

The Future of Social Inequality

This image of recurring crises and potentially worsening economic difficulties is consistent with the classical Marxian view of what should happen as capitalism matures. It will be recalled that Marx saw such developments as part of capitalism's built-in tendency to bring about conditions conducive to its own destruction or transformation over time.

In Wright's early analyses, he adhered to this same general view of capitalism's future. In other words, Wright expected that both the state and the bourgeoisie would eventually be less and less able to ease the repeated economic crises wracking the capitalist system. These conditions, in turn, were likely to increase popular discontent and enhance the prospects for a socialist revolution in future society.

More recently, however, Wright has frankly expressed some doubts about a socialist transformation of capitalism occurring in this way. First of all, Wright is less convinced than in the past about the universal applicability of the traditional Marxist perspective on social change. He does retain the essential Marxian view that human history has a certain underlying tendency or directionality, and is more than a "random walk" of events that are simply "strung together" (Wright et al., 1992: 11). He also believes that Marxism "may still be humanity's best hope for understanding [and changing] the social world" (Wright et al., 1992: ix). Thus, he believes that our hopes for emancipation, not only from class injustice but from problems such as racism and sexism as well, still seem to be closely tied to a socialist future. Even so, Wright now argues that contemporary Marxist theory must proceed "with vastly diminished explanatory pretensions" when compared with the claims of the past (Wright et al., 1992: ix; Wright, 1994: 193; 2005: 4–5). Similarly, he questions some of the central tenets of orthodox Marxism, including the core belief that "capitalism is a transient social form," a system that, in the course of its own development, must become "progressively less viable" (Wright, 1994: 113).

While Wright still holds with Marx's idea that there are inherent contradictions in the capitalist system, he also notes that such contradictions are only necessary conditions for proletarian revolution, and are not sufficient in themselves to ensure the triumph of socialism (1994: 230). Of course, the latter argument is one that Marx also made more than a century ago. In particular, Marx maintained that one of the crucial preconditions of a successful proletarian revolution is that workers, when made conscious of the ills of the capitalist system, must also develop the desire and the will to choose socialism as a better alternative.

On this point, as well, we can see evidence of a change from Wright's earlier analyses. In his past work, Wright seemed confident that, if the proletariat would simply realize that socialism is in their fundamental interests, then the socialist revolution would surely come to pass. In this view, one key reason why socialism had not taken hold in many countries was that workers had been distracted from seeking their revolutionary destiny, because of their preoccupation with more immediate interests such as wages and job security. Although the material deprivations faced by workers might make such distraction understandable, Wright seemed assured that the proletariat would engage in the struggle for socialism, if only its members had a complete comprehension of their exploitation by the capitalist class (1978: 88–90).

It is now apparent that there are significant difficulties with such a view. To begin with, it really assumes, rather than demonstrates, that workers have a fundamental stake in socialism. While such an assumption could be correct, as it stands it is largely a Marxist axiom, a statement that must be true by definition. As a result, the statement ignores other explanations for working-class inaction, including the possibility that workers themselves are capable of judging the merits of socialism and have found them wanting. It is plausible that, on observing the situation in the former Soviet Union, or in contemporary China, or in other instances of socialism in practice, most workers have chosen capitalism as the lesser of two evils, despite its exploitation and injustice. In that event, capitalism would be more suited to both the immediate and the fundamental interests of workers, at least compared with the many failed attempts to establish socialism in the modern era.

In his current writings, Wright largely accepts this point. While it may be true that workers have generally endured or consented to their subordinate position in the capitalist economy, this does not mean that they suffer from "delusions" about the existing system (Wright, 1994: 79). Following Cohen, Wright asserts that workers are not "helpless dupes of bourgeois ideology," who just don't understand the reality of their exploitation and domination by the bourgeoisie (Wright et al., 1992: 44; see Cohen, 1978: 245). On the contrary, it is possible to discern a logic or "strategic rationality" in the choices that workers have made in not pursuing the overthrow of capitalism. Considering the present circumstances confronting workers—the real dangers associated with revolutionary upheaval, the obvious lack of success of socialist systems in the twentieth century, and the relative material advantages for people living in many capitalist societies—it should not be surprising that the proletariat has generally shown little interest in the socialist cause. As Wright puts it, when faced with the choice of socialism over capitalism in the contemporary period, "workers plainly have much more to lose than their chains" (Wright et al., 1992: 41; also Wright, 1994: 79). In fact, sounding somewhat like Dahrendorf and Parkin, Wright now allows for the possibility of "positive class compromises" in present-day capitalism—i.e., the institution of mutually agreed-upon arrangements between capitalists and workers in which the interests of *both* classes are satisfied, at least in the short or medium term (Wright, 2000).

The size of the proletariat is another important issue to consider when assessing the prospects for socialism in the future. Whether we apply Wright's original or his revised typology of classes, it is apparent that only a minority of the population is actually located in the working class proper (1978: 84; 1985: 195; 1989a: 9; 1994: 188). In the United States, for example, Wright has estimated that perhaps 40 to 50 percent of the class structure is proletarian, depending on the criteria for classification and the year in which the estimate is made. Even if all these members of the working class take a strong interest in socialism, a point that has already been questioned, a majority remain whose class interests partially or completely oppose capitalism's demise. Of these, only 1 or 2 percent are in the bourgeoisie, while another 5 or 6 percent are in the petty bourgeoisie. Thus, the prospects for socialist revolution seem to hinge on those who occupy the numerous contradictory locations or other class categories that form close to half of Wright's class structure (Wright et al., 1982).

If all the people in these positions were to resolve their ambivalent interests in favour of the proletariat, the massive support necessary for a revolutionary overthrow of capitalism could be attainable. In some of Wright's earlier research, he perceived signs of "proletarianization" among some of the technical, white-collar, and other employees in these locations, because some have seen their objective conditions of labour come more and more to resemble those of the working class. However, compared with Poulantzas and others, Wright is less certain that all such intermediate positions will become proletarian; instead, it is likely that many will remain distinct for some time to come, and new locations of this sort will continue to be created as technology advances (1979: 28–32; Wright and Singelmann, 1982; see also Gagliani, 1981). Subsequent research by Wright himself suggests that, in the United States, there has been no recent reduction in these intermediate positions. Instead, there has been a significant "expansion of the middle class" in recent decades, even within the traditional petty-bourgeoisie category that both Marxists and non-Marxists have assumed was in decline (Wright and Martin, 1987: 25; Steinmetz and Wright, 1989: 973; see also Wright et al., 1992: 41). Other American research has also shown a growth in highly skilled and highly paid jobs that are typically associated with the middle class, although this trend is combined with a growth among low-level service jobs, as well (Morris, Bernhardt, and Handcock, 1994). A similar rise in skilled or middle-class positions has been reported in Canada for about the same time period (Myles, 1988; Picot, Myles, and Wannell, 1990; see also Hunter, 1988), and also some growth in the petty bourgeoisie (Arai, 1995). The persistence of these categories, coupled with whatever disaffection exists toward socialism in the proletariat itself, makes it less than probable that a majority desire for the overthrow of capitalism is imminent.

Nevertheless, assuming that the working class is eventually able to overcome these and other impediments to achieving revolutionary consciousness, Wright then raises another key concern: that workers be allowed a real part in shaping revolutionary change, especially through the political apparatuses of the state. Like Poulantzas, Wright rejects the old view that successful revolution means a sudden and total smashing of the capitalist state, since this typically leads to a totalitarian socialist regime that excludes workers from power and subverts the very principles it is supposed to uphold (see 1985: 83–84). Wright offers no detailed alternative strategy, but, like Poulantzas, he holds the basic hope that working-class community councils and similar organizations can arise to establish "direct democracy on the fringes of the state administration" (1978: 245). Then, as the state increases its involvement in running the economy and the rest of society, workers will be in a position to apply their developing organizational skills and other "capacities" by participating in state programs directly. In this way, workers can ultimately achieve their rightful majority influence through democratic means.

Still, as Wright acknowledges, this smooth and democratic achievement of a society ruled by all workers is likely to encounter considerable opposition, depending on whether the state's repressive branches—the police and military—are used against it, and on whether possible economic reprisals by the bourgeoisie—such as an "investment strike" involving the flight of businesses out of the country—are successful in impeding its course (1978: 250–251). Apart from these potential obstacles, however, the final hurdle for the universal society of workers may well be organized socialist political parties themselves. To Wright, these are a necessary focal point for concerted proletarian action, and a successful transformation of the capitalist state from within seems inconceivable without them. Even so, they are subject to the same pressures for administrative hierarchy and bureaucratic control as any other organization and, as a result, threaten to undermine the broadly based worker involvement in planning and running society that socialism stands for (1978: 247, 252; 1985: 84–85; 1994: 155–156).

On this final point, then, Wright comes down closer to Weber than many other Marxists do, for he suggests that even if workers actively seek the socialist transformation, and even if it succeeds, organizational exigencies persist that make a truly democratic socialism difficult (1978: 216, 225). However, in keeping with Marx, Poulantzas, and most other Marxist thinkers, it appears that Wright would rather risk the future on the possibility that workers can overcome these hurdles than on Weber's hope that some enlightened political elite can save capitalism from bureaucratic abuses and its own internal contradictions. Workers must develop and retain the capacity to rule themselves if human societies are to progress, and the chances of this occurring are, in the Marxist view, clearly better under socialism than under any form of capitalism.

Of course, given the many problems encountered by past and present socialist regimes in the twentieth century, we might wonder that Wright and other scholars on the political left would still retain their belief in the advantages of socialism and their commitment to the Marxist vision. In fact, as Wright himself has stated, many Marxists have either abandoned the perspective altogether, or else have sought out some "post-Marxist" transcendence of their old ideas (e.g., Wright et al., 1992: 1–2; Wright, 1994: 12–13). Still, Wright is optimistic that the current period of questioning and uncertainty facing Marxism is not so much a "crisis" as it is an opportunity for rethinking and revitalization, for "growing pains" and continuing "maturation." In spite of the acknowledged problems in traditional Marxist analysis, Wright's position is that any assumptions about the demise of both socialism and Marxism are decidedly premature. Rather than see the socialist project as dead and gone, we should instead view it as an uncompleted work in progress. To Wright, then, Marxist theory, in renewed form, is as relevant as before, and Marx's hope of achieving universal human emancipation through socialist revolution is still a goal that is well worth pursuing (see Wright et al., 1992: 1–2, 190–191; Wright, 1994: 254–255; 2005: 30, 191).

Summary Observations

In this section, we have focused on Erik Olin Wright's attempt to reorient traditional Marxism to the analysis of contemporary class structures, the capitalist state, and other topics relevant to the study of social inequality. Overall, Wright's discussion is more clearly conceived and internally consistent than Poulantzas's neo-Marxist formulation, especially with respect to the problem of class structure. While both Poulantzas and Wright provide noteworthy treatments of the state, a topic largely ignored by many early Marxists, here too Wright's analysis seems preferable to Poulantzas's rather obscure account.

Wright does not offer a comprehensive treatment of the concept of power, but his recent writings suggest an increasing acknowledgment of the significance of noneconomic forms of power or domination in social structures. Wright's current work also devotes more attention to examining the problems of gender, race, and other nonclass forms of inequality. Thus, although his principal concern is still with the analysis of class structure and the continued development of Marxist theory, many of Wright's insights are notable for their contributions to a more general understanding of social inequality, in all its forms and manifestations.

Finally, in his views about the future, Wright takes a more guarded and qualified position on the prospects for a socialist revolution than he once did. In particular, Wright recognizes that, in the contemporary period, there are good reasons why the working class might dispute the claim that a socialist future is in their fundamental interests. Here and elsewhere, Wright exhibits a general willingness to question many of the standard assumptions in Marxist analysis. In fact, his proclivity for theoretical rethinking and reconstruction has led some traditional Marxist scholars to criticize him for being too malleable. Nevertheless, other observers will argue that this flexibility and receptiveness to new insights are two of the main strengths of Wright's current "analytical" approach to Marxism. There are certainly many non-Marxist theorists who would applaud this approach, although some might also question how Marxist it really is in the end. Such questions provide the backdrop for the next perspective to be considered: Parkin's neo-Weberian critique of Marxism.

FRANK PARKIN

The Bourgeois Critique of Marxism

Frank Parkin's work is perhaps the clearest contemporary attempt by an avowed neo-Weberian to provide a renewed perspective on social inequality (1979, 1983). In his analyses, Parkin dissociates himself from certain early interpreters of Weber, especially those structural function-alists who erased from Weber's original analysis his important affinities with Marx (Parkin, 1972: 17–18; 1978: 602–604; 1979: 48). At the same time, however, Parkin is highly critical of most Marxists as well, because of their inordinate preoccupation with one aspect of social inequality: capitalist economic production and the division between bourgeoisie and proletariat arising within it. Parkin's prime purpose is to act as a non-Marxist or "bourgeois" sociologist and criticize the flaws arising from this narrow focus. His own position is that there are other important class cleavages in modern society to consider, as well as numerous other bases for exploitation that are distinct from class and that persist independently of the class structure itself. These other types of exploitation involve a range of social criteria that vary in importance across different societies but typically include race, ethnicity, gender, and religion, among others (1979: 9, 89).

To Parkin, it is instructive that whenever Marxists such as Poulantzas and Wright try to incorporate such complications into the conventional two-class model of Marxism, they produce perspectives that invariably resemble Weber's in key respects. This prompts Parkin to note that "inside every neo-Marxist there seems to be a Weberian struggling to get out" (1979: 25; see also Wright, 1994: 92). Wright is one leading Marxist scholar who acknowledges these overlaps between recent Marxist and Weberian thought, although it is interesting that Wright also responds to Parkin's gibe by suggesting the possibility that "inside every left-wing Weberian there is a Marxist struggling to stay hidden" (Wright, 1997: 35; 2005: 27).

In either event, it is Parkin's conclusion that all manifestations of structured inequality can and should be examined using a single, essen-tially Weberian, conceptual framework. This unified scheme treats power relations, not class relations, as the elemental factor generating inequality,

Frank Parkin, born 1940

© Jerry Bauer.

but elaborates Weber's original notion of power by linking it to his related though less familiar idea of *social closure*. Because of its central role in Parkin's analysis, we begin with an assessment of his discussion of power as social closure. We then consider Parkin's treatment of class, the modern state, and the outlook for inequality in future society.

Power and Social Closure

Parkin concurs with the view put forward in Chapter 3 that, for Weber, the overall structure of inequality in society stems from a general and continuing struggle for power (1972: 46; 1979: 44). According to Parkin, however, it is unclear in Weber's discussion precisely what the source or location of power is in society, particularly because Weber's definition of the term is never completely satisfactory (1979: 46). Parkin's remedy for this problem is to combine the idea of power with social closure, a less prominent concept in Weber that refers to the various processes by which some social groupings restrict others from "access to resources and opportunities." From this viewpoint, power is really a "built-in attribute" of any closure situation, denoting one's degree of access to these resources and opportunities (1979: 44–46; see Weber, 1922: 342).

Parkin suggests two basic forms that social closure can assume: *exclusion*, which is the prime means by which dominant factions deny power to subordinates, and *usurpation*, which is the key means by which subordinates seek to wrest at least some power back from those who dominate them. Of the two, exclusion is by far the more effective form in modern societies, for it is largely established in legal rules and regulations that enjoy the official sanction of the state's justice system and, if necessary, its repressive agencies as well. The obvious example of exclusion in capitalism is the legal right to own private property, which excludes workers from power over the production process. As for usurpation, it is a secondary process for the most part, through which those subordinates denied power by formal exclusion attempt to consolidate themselves and mobilize power in an "upward direction." In its most extreme guise, usurpation could mean a complete overthrow of the ruling faction, as in a proletarian revolution to oust the bourgeoisie from power. Typically, however, usurpation involves more moderate and less potent kinds of action, the most common being a collective withdrawal or disruption of services through strikes or demonstrations. In contrast to exclusion, such usurpationary acts are less often given formal recognition or protection by the state. Such tactics are frequently accorded only grudging or partial legitimation and may even be outlawed in some circumstances. An example of the latter would be to disallow the formation of unions in occupations that are deemed to provide essential services to the public. Thus, like Lenski before him, Parkin envisions a two-way flow of power in social hierarchies, but one that is asymmetrical, in that the exclusionary powers of the dominant faction generally override the usurpationary responses of subordinates (1979: 45, 58, 74, 98).

So far, we have used examples of class relationships to illustrate Parkin's two forms of social closure. However, Parkin stresses that closure processes are the common factor behind all structures of inequality, including ethnic, religious, sexual, and other types of exploitation in addition to class relations. Once again, Parkin is critical of Marxist analysts here, most of whom ignore these other types or simply give them passing mention as phenomena that obscure from view the "real" struggle between classes. But, to Parkin, the struggles between blacks and whites in the old South Africa, Catholics and Protestants in Northern Ireland, or women and men in most every nation are all conflicts over closure, over access to resources and opportunities. Moreover, such antagonisms and the inequalities connected with them occur in both capitalist and socialist societies, regardless of whether class inequality exists or not. Therefore, it is incorrect to assume that these nonclass inequalities are somehow secondary to or derivative from class struggle (1979: 113–114). On the contrary, these other bases for inequality are frequently more important than class in explaining social change or collective action in some countries (1978: 621–622). What is required, then, is a recognition of these other clashes between interest groups and an analysis of how they reinforce, negate, or otherwise interact with class closure. Although we have seen that Wright is one Marxist who pays greater heed to these nonclass forms of social closure, Parkin would maintain that this is not typical of neo-Marxist theorists in general.

Closure and Class Structure

Having noted the essential process of social closure and the various systems of power and inequality it can engender, Parkin gives particular attention to the problem of class structure. Despite his criticisms of Marxist analysis, Parkin clearly holds with Marx, as well as Weber, in emphasizing the pivotal role of property relations in defining classes (see also Murphy, 1988: 26). In fact, control of productive property continues to be "the most important form of social closure" in society, since it can mean exclusion from access to the material means of survival itself (1979: 53). However, following Weber, Parkin notes that the class system is increasingly being shaped by a second key type of exclusion: the use of formal "credentials," especially educational certification, to close off privileged positions from others (1979: 54; Weber, 1922: 344; see also Collins, 1979). Particularly in such fields as medicine and law, a few incumbents have gained licence from the state to special forms of knowledge and practice, resulting in a legal monopoly over professional services. Parkin believes these credentials are so important that those who hold them are the second layer of the dominant class, just below those who have exclusive control of productive property (1979: 57–58). Parkin's inclusion of educated professionals within the dominant class may be questionable, to the extent that their influence in society would seem clearly to be less significant than that of the most powerful capitalists, for example. Nevertheless, Parkin is correct in recognizing the great advantages that highly exclusive credentials can give to those who possess them. As we have already discussed, current Marxist scholars like Wright now also lay considerably more stress than in the past on the significance of credentials or recognized skills for determining class location.

Outside the dominant class, the class structure in Parkin's view seems to form a graded hierarchy of people who have varying degrees of usurpationary influence, occasionally combined with a partial capacity for the exclusion of others. Workers without property or credentials form the subordinate class, although they vary internally according to whether or not they enjoy usurpationary influence because of union affiliation, and also according to the strength or "disruptive potential" of their union relative to others (1979: 80, 93). Between the dominant and subordinate classes, Parkin identifies a range of "intermediate groups" who exercise incomplete forms of both exclusion and usurpation, in what Parkin calls "dual closure." The lower end of this middle range includes skilled tradespeople, who usually are unionized and can also invoke limited exclusionary closure through apprenticeship systems and other credential mechanisms. These forms of certification are less exclusive than those given doctors and other professions but still provide workers who have such "papers" some advantages over workers who do not. The remaining intermediate positions primarily involve white-collar "semiprofessions," a mixture of teachers, nurses, social workers, and the like, who may shade into the more established and more exclusive professions at the upper end but normally fall short of the complete exclusionary closure that doctors and lawyers have typically attained. Such semiprofessionals are often state employees, who lack a true monopoly over the knowledge or services they dispense, and who may thus resort to unionization and other usurpationary tactics to defend or improve their position in the collective struggle of interest groups (1979: 56–57, 102–108).

In total, then, Parkin's portrayal of the class system involves a dominant class of people with exclusionary rights to property and to special credentials, a subordinate class of workers with only varying amounts of usurpationary power, and a middle range of semiprofessionals and skilled workers with different mixes of both types of closure at their disposal (see also Parkin, 1983). Despite his emphasis on closure, Parkin observes that the class structure is open or permeable to some extent. The dominant faction in modern liberal democracies harbours a degree of ideological commitment to individual opportunity, so that some with talent will be allowed and even encouraged to move up from humble origins, while others of privileged background who lack ability must eventually move down. Nevertheless, this sifting process does not alter the closure principles themselves and falls far short of negating the advantages of those who inherit property and other rights of closure. In addition, any commitment to individual opportunity is belied by the tendency of collective or ascriptive criteria, such as race or gender, to become interwoven with considerations of merit and performance. Thus, blacks or women could be denied entry to the dominant class irrespective of their personal qualifications or abilities and could be relegated to the lowest reaches of the subordinate class by white male workers, who may jealously guard what few prerogatives they have themselves (1979: 63–68, 90–91). These latter possibilities suggest some of the ways in which the other bases of social closure are embedded in the class structure and can interact with it (see Murphy, 1988: 80–81; see also Tilly, 1998). These nonclass bases also illustrate that, in Parkin's view, the idea of social closure provides us with a more powerful conceptual tool for explaining social inequality in general than does the idea of class.

The Role of the State

Parkin considers the upsurge of interest in the state to be one of the most novel features of contemporary discussions of social inequality (1978:

617). His own emphasis on exclusionary closure, established and upheld by the legal and repressive branches of the state, suggests that Parkin also sees the state as crucial to contemporary systems of inequality. However, Parkin contends that the concept of the state has been misused or misunderstood by many theorists, especially those on the left. There is an apparent inconsistency in those Marxists such as Poulantzas who argue for the state's relative autonomy, as if it were a separate force in society, and yet claim that it has no power of its own (Parkin, 1978: 618; see Miliband, 1969). Given Parkin's Weberian leanings, it may be surprising that his own view of state power is generally close to that of Poulantzas. That is, Parkin suggests it is not really the state that has power; it is *people* who have power *through* the state, especially those people who are most able to infiltrate state positions or influence state personnel from the outside. This means that the state itself is mainly "an instrument of social domination"—in fact, an elaborate cluster of such instruments that subsumes the "administrative, judicial, military, and coercive institutions" of society. These structures are subject to separate and even contradictory manipulation by the various antagonists in the struggle for social closure. Using a different metaphor, Parkin also compares the state to a "mirror" that reflects the overall outcome of conflicts involving classes, races, genders, and the other key interest groups in society (1979: 138–139; see also 1978: 619).

Nevertheless, despite this image of a complex struggle or contest to acquire power through the state, Parkin also argues that a few dominant factions, particularly the dominant economic class, are typically able to gain majority control over the means of social closure (1972: 181–182). In addition, while Parkin apparently believes that the state is without power of its own, he seems to allow for one crucial exception: where a single political party is able to take complete control of the state, as occurs in state-socialist systems, for example. In these situations, the will of a single political faction becomes "fused" with the entire state apparatus into one totalitarian, omnipotent "party-state." Under socialism, this fusion means the end of bourgeois class domination, but it can also mean the centralization of all power within the party-state (1979: 140). The serious problems that Parkin perceives in such a system provide the basis for his assessment of the future of inequality and the relative merits of capitalism, socialism, and social democracy.

Capitalism, Socialism, and Social Democracy

Notwithstanding the acknowledged exploitation, injustice, and other flaws in capitalism, Parkin is not surprised that workers in most developed countries appear to find the prospect of socialism even less palatable. To Parkin, the distinct lack of interest in the socialist cause in many capitalist countries is less attributable to "false consciousness" or "mystification" within the working class than to the regrettable examples provided by the former Soviet Union and other attempts to create state-socialist systems. Parkin offers at least four problems that can be identified in contemporary socialist regimes, several of which are now recognized by recent Marxist writers, as well.

First, although organized repression is found in all societies to some extent, the fusion of party and state within totalitarian socialist regimes on the left (and, of course, totalitarian fascist regimes on the right) seems far more conducive to the sustained use of violence and terror than is typical of Western capitalist societies. It is partly because of this spectre of violence on the left, of Stalinist purges and "the possibility that the Red Army might be mobilized for other than purely defensive purposes," that workers in other nations have been suspicious of putting socialist doctrines into practice (1979: 201).

A second impediment to popular support for socialism is the evidence that early Marxist revolutionaries, including Stalin, Lenin, Lukacs, and others, placed much greater faith in an elite "vanguard" party than in a general groundswell of workers' involvement in shaping and directing socialist transformation. It may be that Lenin and others expected this dominance by the elite to be only temporary. In addition, as we have seen, many modern Marxists such as Poulantzas and Wright strongly advocate broad proletarian participation in any future socialist initiatives. Nevertheless, the impression among many workers—that the dictatorship of the proletariat is just a euphemism for totalitarian control by the party-state—is partly responsible for their disavowal of socialist revolution. Even minimal freedoms under capitalism—such as the right of political dissent or the choice of more than one political party to vote for—imply, by comparison, a much more democratic system (1979: 153–155, 178–182).

A third factor acting against the success of socialism is the perception among many workers that the elimination of private control over production has not eradicated inequalities in power or in the distribution of resources and opportunities. Removal of property rights under socialism has simply meant that other forms of exclusionary closure have become the salient bases for inequality. One key example involves the exclusion of the mass of the population from the elite category of intellectuals, scientists, officials, and bureaucrats, who have special educational credentials or strategic positions in the party hierarchy with which to exact special privileges.

Finally, although the differences between top and bottom are probably smaller under state socialism, workers in capitalist countries still have the perception that there is less wealth to spread around in socialist economies, so that being more equal under socialism may still mean having a more meagre material existence than under capitalism (Parkin, 1979: 185–187). If this situation prevails or is perceived to prevail, and if workers put greater stress on distributive issues than some Marxists would prefer, it is unlikely that state socialism can serve as a sufficiently attractive replacement for capitalism.

Although Parkin's assessment implies that neither capitalism nor socialism offers a particularly rosy future, there is at least one other possibility to consider, one that Parkin himself appears to support. Increasingly, in his view, the real choice has become one of state socialism on the one hand and *social democracy,* not pure capitalism, on the other. In the social democracies of Scandinavia, for example, capitalists can still retain private control of production and the exclusionary powers that go with it. However, the tensions generated between capitalists and workers can be largely "contained," though never eliminated, through state legislation and other means by which workers get enough usurpationary power that the bourgeoisie retains dominance, but just barely. The trick is to find a balancing point that leaves the capitalist class with sufficient incentive to invest and accumulate surplus but that simultaneously reduces the exploitation of workers to the lowest possible level (1979: 189). This containment of class struggle need not mean that Dahrendorf's complete legitimation or institutionalization of class conflict is achieved, only that workers and capitalists alike are able to decide how far they can or should go in usurping privilege or excluding others from it.

It is interesting that this social-democratic solution to the problem of inequality returns us to certain basic assumptions about human nature. Social democracy in these terms locates itself somewhere between Marx's belief in the potential selflessness of humanity and the conservative or structural-functionalist insistence on people's inherent motivation to acquire more power or rewards than others. Like Marxism, Parkin's version of social democracy assumes that all people can flourish under more egalitarian conditions and that, in the proper setting, people are capable of tying personal interest more closely to a concern for the collective good. At the same time, this approach also adopts the argument, made by many conservative thinkers, that no amount of structural change can totally eliminate the "small inner core of human individuality," with its self-interested motives and its belief in differential rewards for differential talent or effort (Parkin, 1979: 189–190). From this perspective, some degree of inequality, restricted and reduced though it may be, is perhaps a natural, inevitable, or necessary outcome of social relationships.

Of course, in posing this alternative to state socialism, Parkin is well aware that social democracy has also run into snags when its principles are put into practice. For example, he argues that the record of social-democratic governments in western Europe and elsewhere has not always been stellar in reducing material inequalities or the concentrated power of capital (1979: 200). Yet even modest advances along these lines, when compared with the militarism, totalitarianism, and material shortages of socialist societies, would probably make some form of social democracy the preferred system for most of the working population.

Summary Observations

Parkin's neo-Weberian perspective on social inequality has been the subject in this section of the chapter. Overall, his approach represents a provocative and constructive attempt to reorient Weber's original power perspective by introducing into prominence the concept of social closure. Parkin is able to sustain both Marx's and Weber's stress on class as the crux of social inequality, while also accounting for gender, race, and other important bases for closure or power relations that regularly emerge and become established in social settings. While Parkin's approach thus contributes significantly to a truly general conception of social inequality, there are at least two points that would benefit from additional elaboration or reconsideration.

First, Parkin's main focus on the legal bases for power or closure, though justifiable, could be expanded to include a more detailed discussion of the nonlegal bases for social domination arising from informal, but equally effective, rules established in traditional beliefs, customs, and habitual practices. Though Parkin is clearly aware of these forms of closure, and the manner in which they can exclude women or racial minorities from power in spite of their *legal* equality, it would be useful to give them greater prominence, especially since this would be consistent with Weber's original analysis.

Second, Parkin's suggestion that the state has no power, except in such situations as the fusion of party and state under state socialism, seems clearly to be open to question (see Murphy, 1988: 56, 113). Given Parkin's neo-Weberian viewpoint, it stands to reason that he would follow Weber in noting the special power attached to state bureaucrats, power that is not reducible to the class, ethnicity, gender, party, or other external affiliations they possess but that inheres in their roles as bureaucratic administrators and decision makers internal to the state itself. In that sense, heads of state organizations have power in the same way that any other privileged factions do. In a similar vein, one could note that the unique coercive powers of the state's military and police branches also give them special and, at times, even ultimate control in society.

ANTHONY GIDDENS
The Structuration of Class, Power, and Inequality

The next major approach to social inequality we shall examine is that found in the writings of Anthony Giddens. Giddens is recognized as one of the leading figures in contemporary sociological thought. It has been claimed, in fact, that among social theorists Giddens is "certainly the pre-eminent figure in the English-speaking world" (Cassell, 1993: 1). Giddens's most notable contributions include his insightful analyses of Marx, Weber, Durkheim, and other classical theorists (e.g., 1971, 1972, 1986), as well as his extensive efforts to devise his own fundamental reorientation of general sociological theory (e.g., Giddens, 1976, 1977, 1979, 1981b, 1984, 1985, 1990, 1994). In the process of pursuing these and other projects, Giddens has reconsidered several topics of special relevance to the study of social inequality, including past and present conceptions of class, power, and the capitalist state (1973, 1981a, 1981b, 1985; see also Giddens, 2003; Giddens and Diamond, 2005).

It was noted at the beginning of this chapter that Giddens is often categorized as a neo-Weberian, particularly by some Marxist critics of his views (Binns, 1977; Crompton and Gubbay, 1977; Wright, 1978, 1979; Burris, 1987; but see Wright et al., 1992: 61). While this assessment seems appropriate at a general level of discussion, it should be observed that, as Giddens himself argues, on some issues he owes rather more to Marx than to Weber. Moreover, in generating the finer points of his own perspective, Giddens frequently draws on and builds from a wide range of additional viewpoints. These conditions should be kept in mind when noting Giddens's location between the Marxist and Weberian camps on our left–right continuum in Figure 6.1 (page 117). Giddens has recently argued that it is time to go beyond the old debates between left and right that informed sociological analysis in the past (Giddens, 1994). At the same time, he has also advocated the need for a neo-progressive "third way," an approach that in some senses lies between, but actually transcends, the old-style socialism of the traditional left and the largely unregulated capitalism of the conservative right (Giddens, 1998, 2000a, 2002, 2003).

In the present context, then, it is appropriate to place Giddens where he is on the continuum because of his joint sympathies with basic ideas from both Marx and Weber, who, as we have seen, have more in common than some analysts realize or acknowledge. Giddens's position on the continuum also signifies his felt need to supplement and modify the views of these classical thinkers. Giddens concurs with Weber's initial attempt to deal with certain points left largely undeveloped in Marx, especially the pluralist nature of class and power, the importance of bureaucracy in modern systems of domination, and the role of the state as the focus of legal and repressive power in advanced societies. In turn, Giddens has emphasized the need to correct and elaborate the manner in which Weber has dealt with these and other issues (1981a: 296–300; see also 1973: 100–102; 1984: xxxvi).

One of the outcomes of Giddens's efforts to expand and redirect existing perspectives is his own "structuration" approach to the analysis of social processes. It is beyond our purposes to explore all the intricacies of this general orientation; nevertheless, its specific application to the

Anthony Giddens,
born 1938

Reprinted by permission of Anthony
Giddens.

problem of social inequality is crucial to consider because, in some respects, Giddens offers one of the most promising strategies for developing a comprehensive theoretical approach to this area of study (see Turner, 1986: 974–975; Sewell, 1992: 4). As with the other theorists we have examined, the discussion of Giddens centres on how he deals with four main issues: the concept of class or class structure, the significance of power in systems of social inequality, the role of the state, and the prospects for inequality in future society.

Class Structure in Advanced Societies

The beginnings of Giddens's overall structuration perspective can be found in his initial analysis of the class structure in advanced societies (1973; 1981a). Like Wright, Giddens argues in his early work that the two-class system envisioned by Marx, involving bourgeoisie and proletariat, is acceptable primarily as an "abstract" or "pure" model of the capitalist class structure, one that omits certain residual class locations in real societies (1973: 27–28). Giddens agrees with all Marxists and most Weberians that the crucial factor generating this class system is "ownership or exclusion from ownership of property in the means of production" (1973: 100, 271–272). In the end, however, his links to Weber on this issue seem clearer, for he adopts the original Weberian position that classes are largely products of differences in power among groups within the capitalist market. In the economic setting, capitalists enjoy greater power than workers because they retain rights over productive property, while workers have only the right to sell their labour in exchange for a living wage paid by capitalists. Where Giddens reveals his central tie to Weber, and his basic divergence from Marx, is in his contention that there is a third important category of resources that underlies the different economic power, or "market capacity," of people under capitalism. This third factor is the "possession of recognized skills" and "educational qualifications" (1973: 101–103).

Using this mixture of Marxian and Weberian precepts, Giddens argues that the three rights or capacities—property, education or skills, and labour power—are fundamental to a corresponding "threefold class structure" that is "generic to capitalist society." It is primarily because these three bases of power predominate in the economic sphere that social relationships arise among an "upper" class of those who control most productive property, a "middle" class of those without appreciable property who nonetheless have special education or skills to exchange in the market, and a "lower" or "working" class who normally have only their labour power to sell.

Of course, if this three-class scheme is taken too literally, it is subject to the same criticisms as Marx's two-class model, since there are many exceptions that do not fit easily into any of the categories. Giddens himself notes that one cannot draw absolutely clear boundaries between classes, as if they were "lines on a map" (1973: 273). This is because some groupings, such as the old petty bourgeoisie or independently employed doctors and other educated professionals, tend to straddle class lines, in that they have partial access to more than one of the three types of rights or capacities. Using terminology similar to that of Parkin, Giddens suggests that the degree of "closure" or "exclusion" produced by these rights is not always complete (e.g., 1973: 107). Besides, the extent to which such mixed positions actually blur class lines will vary across different societies, and even different regions within the same society, so that no single model can capture precisely all the detailed differences in class structure that arise in this range of situations.

It is here that Giddens introduces the idea of *structuration,* primarily as an aid in dealing with these anomalies in real class systems. Rather than speak of classes as if they were discrete groups, explicitly delineated and separated in all instances, Giddens recommends that class structure be construed as a *variable* phenomenon that is generally anchored in a three-class system but that differs in its *degree* of structuration, in the extent to which classes are generated and reproduced over time and place as identifiable, distinct social clusters. In this sense, one can argue that property, educational qualifications, and labour power do act as the major powers or rights that interconnect, or *mediate* between, the economy and the classes arising from it. The three resulting classes will be more clearly defined or structurated in those societies or situations where these three mediate factors tend not to be mixed together in the same occupations or positions (see also Wright, 1985: 106, 112).

In addition, at a *proximate* level, or at the level of everyday life experience, Giddens suggests three other factors that can either blur or reinforce the three-class model he outlines. The first proximate factor is the way in which labour is divided in the work setting itself. In most societies, for example, manual labourers are physically separated from specially trained or educated nonmanual workers and almost always perform different tasks. This separation discourages interaction between the two types of workers and reduces the likelihood that they will share much in common with each other. As a result, the generic tendency toward a split between the middle and working classes, which arises out of their different market capacities, is reinforced still further. In contrast, we can conceive of a situation where there is little or no physical separation of the two types of workers in the workplace, or where employees are not distinguished along manual and nonmanual lines because people regularly rotate in and out of both kinds of jobs, as can happen in some small firms, for example. In societies where these patterns predominate, class divisions would be relatively weaker and class structuration less pronounced.

A second proximate factor to consider is the manner in which authority relations are structured in the work setting. Sometimes there is no real difference in the decision-making or supervisory powers of manual workers and specially trained personnel, while in other situations special personnel have authority prerogatives that divide them from manual workers in the same way that their greater education or training does. In the former case, there would be some blurring of the boundary between the middle and working classes, at least on this basis, while in the latter instance the boundary would be enhanced.

The final proximate factor in class structuration is the pattern of "distributive groupings," by which Giddens means primarily the amount of clustering that occurs because of distinct lifestyles or material consumption habits. For instance, it is possible in some societies to see clear class differences in people's manner of dress, as illustrated in the distinction that is sometimes made between the "white-collar" middle class and the "blue-collar" working class. In contrast, we can imagine a society in which class structuration is not encouraged by such visible distinctions, as in Communist China's early attempts to have all people wear similar clothing, for example. Perhaps the best illustration of distributive groupings that Giddens offers concerns the purchase of housing and the physical segregation or mixing of classes that can result. In societies where the upper, middle, and working classes invariably live in clearly separated areas that do not overlap, their pattern of consumption of housing would obviously reinforce the underlying three-class system and make it more readily discernible as a social *reality*. On the other hand, if the predominant pattern in a society or region is a mixing of diverse people in the same neighbourhood, regardless of their market capacities in the economic sphere, then class structuration would be less pronounced and class lines would be less clearly defined than otherwise (1973: 107–110).

What Giddens offers, then, is a basic three-class model that differs from Marx's dichotomous view in two key ways: in its designation of a middle class of educated and skilled personnel, who tend to differ from both capitalists and manual workers; and in its incorporation of a variable element in the conception of class structure, which allows for the possibility that classes in different settings can be more or less distinctly delineated or structurated, depending on the extent to which these six particular factors act in unison, or against one another, in promoting clear class cleavages. Though there are significant differences in both terminology and focus, Giddens thus shares with many recent theorists a concern with complications that, in contrast to Marx's pure two-class model, are still present in existing capitalist societies.

Perhaps the most telling aspect of Giddens's analysis is his conclusion that these complexities do not negate the truths in Marx but are more significant for understanding capitalist class systems than some current Marxists believe. Hence, the middle class is likely to be a persistent reality in advanced capitalism, not some transient fraction or secondary set of class locations. In addition, class affiliations will continue to be elaborate and complex at times, not only because of the variable mix of mediate and proximate divisions, but also because factors such as gender and ethnicity interact with class structuration (1973: 111–112; 1981a: 304–307). These views underscore the pluralist image of class and inequality that, more than anything else perhaps, separates Giddens from traditional Marxism (1973: 273–274). We now consider how this pluralism in Giddens carries over into his pivotal analysis of power.

Power and Domination

For some time, one of Giddens's central concerns has involved refining the concept of power and incorporating it into his more general theory of structuration (1979, 1981b, 1984, 1985). In characteristic fashion, Giddens addresses this problem by noting Weber's advances over Marx, who never attempted a complete analysis of power, while at the same time claiming that Weber's view of power is itself in need of elaboration and revision (1981b: 4–5).

We shall avoid a detailed delineation of Giddens's treatment of power, since it involves more subtle distinctions and more complex terminology than are appropriate for our purposes. However, in more simplified form, Giddens's formulation is useful here because it connects rather closely with the Weberian view that still prevails on this issue, and also because it adds to Weber's conception in certain important respects.

Giddens asserts that power relations are not the only factors linking people together in society but that they do form one such link and are basic to *all* social interaction (1984: 32, 227). Giddens believes, as well, that power, conceived along the lines he suggests, has "universal applicability" to the study of exploitation and inequality (1981b: 201). He defines power as any "relations of autonomy and dependence between actors in which these actors draw upon and reproduce structural properties of *domination*" (1981b: 28–29; 1984: 258). Giddens's use of the term domination to define power shows the close connection that he perceives between these two ideas. It is interesting that Weber also ties power and domination together, as we saw in Chapter 3. The interpretation offered there was that power refers to the capacity of individual actors (or factions) to exercise their will relative to other actors, while domination denotes the regular patterns or structured relations between actors that arise as such power differences are established, routinized, and regenerated over time (see Giddens, 1985: 7–12).

Now, while there are undeniable divergences between Giddens and Weber on other points, they share a rough correspondence in this distinction between power as human capacity and domination as a structural manifestation of power. Thus, Giddens sees domination as "structured asymmetries of resources drawn upon and reconstituted in such power relations" (1981b: 50; see also 1979: 91–93). What Giddens adds to Weber here is a more explicit and sophisticated elucidation of this two-sided sense of power, something that is implicit in Weber but must be extrapolated from his discussion. In fact, Giddens devotes considerable effort to demonstrating that power is "doubled-edged" in a variety of other ways, as well: it typically combines some amount of both repression from above and legitimate compliance from below; it can be used to coerce and constrain but also to liberate and transform; it can operate through formal rules and laws or informal customs and traditions; it can involve acts of commission by superiors but also acts of omission or passive resistance by subordinates; and so on (1979: 88–94; 1981b: 49–51; 1984: 257; 1985: 10–11). Of course, as we have seen, some of the same points can be found in Weber's writings and also in the more recent analyses of Lenski and Parkin, among others. But Giddens is perhaps most notable for his systematic treatment of this duality in power systems and for his recognition that power exists both as a capacity of persons and as a pattern of relations, sustained or reproduced across different times and locations (1984: 111–112, 1985: 7; see also Sewell, 1992; Lemert, 1995).

It is this same reproductive process, whereby people interact in patterned relations so as to establish (or else change) those relations, that is the essence of what Giddens means by structuration (e.g., 1989: 252–259, 300). We have already noted one illustration of this process in the case of class structuration. Class distinctions tend to be solidified or reinforced when people's interactions are determined by their economic capacities (the three mediate factors in capitalism) and by immediate social circumstances related to those capacities (the three proximate factors). The more clearly these bases for interaction separate clusters of actors from one another, the greater class structuration there is.

The class system also provides one of the foremost examples of how power differences and structures of domination develop in social systems. However, in Giddens's view, capitalist class relations are not the only important case of domination or exploitation in society: "Certain fundamental forms of exploitation do not originate with capitalism, or with class divisions more generally" and "not all forms of exploitation can be explained in terms of class domination" (1981b:25, 60). Domination and exploitation also occur in other asymmetrical relations, including those between nation-states—between colonies and imperialistic countries, for example—between ethnic groups, and between the sexes (1981b: 242; 1985: 256; 1989: 265; see also 1994: 242–243). In other words, there is a full range of significant social inequalities that can be identified in modern times—not only class inequalities, but those involving gender, ethnicity, international relations, and so on—and all of these can be understood as manifestations of power differences and the exploitation these differences usually entail. More recently, Giddens has added the problem of generational or age inequality, "the battle between parents and children," to his list of important power struggles. He also contends that, while class divisions are still important, "the conflict of the sexes" increasingly has become the most profound in its implications for social inequality in the present day (Giddens, 1994: 188–189; see also 2000b: 70, 72, 83).

Thus, Giddens leaves no doubt that he has a pluralist view of power. This pluralism is also evident in his overall conception of societal institutions. Although Giddens is highly critical of structural functionalism, and even argues that the term function should be banned from sociological discourse, he nonetheless accepts that the functionalist perspective is correct to recognize the major institutions of society and their embodiment in large-scale social structures (1979: 97; 1981b: 16; 1989: 259–262; for discussion, see Wright et al., 1992: 62–67). As was noted in Chapter 5, structural functionalists conceive of institutions as systems of persistent rules, norms, and values that people tend to live by, or keep in mind, in their interactions with others. These institutions give rise to concrete social structures that roughly correspond to the institutions. An illustration is the formal establishment of a religious structure, or church, in accordance with a particular set of religious values and beliefs.

In a somewhat similar fashion, Giddens delineates four basic types of institutions and connects them to a range of concrete social structures. *Political* institutions operate primarily in the political structures of the state and are concerned with "authorization" or the domination of *people*. *Economic* institutions operate mainly in the economic structure and involve "allocation" or the domination of *material phenomena*. Here Giddens suggests that command over people and command over material things are the two key means for establishing power or domination in society (1981b: 47; 1984: 3; see also 1985: 7–16; 1989: 287; 1994: 132). Giddens's third category includes *symbolic* institutions, those embodied in religion, education, and the communications media, for example, or in what Poulantzas and Wright would call ideological apparatuses. In his final category, Giddens chooses to distinguish *legal/repressive* institutions from the other three types, although their obvious connections to the legal, military, and police branches of the state suggest they could be included more simply under the political category (1981b: 47; 1985: 19; see also Mann, 1986: 11; 1993: 7–10; Runciman, 1989: 14).

In any case, Giddens's discussion of institutions and structures provides further confirmation of his pluralist image of power and society (see also Giddens, 1990: 55–63). This is most obvious, perhaps, in his distinction between political power and economic power, between domination of people and domination of things. In addition, however, the correspondence between institutions and structures is not a simple one-to-one relationship, so that both types of domination operate, at least in secondary form, in the religious and other structures of society as well. It is in this sense that Giddens sees power as a dispersed phenomenon, an integral element of all social life (1981b: 28, 49). This does not mean that power can never be disproportionately concentrated in certain structures or factions. On the contrary, it is clear that those who control the economic and political systems are typically the principal players in the overall power struggle. But it is also the case that any conception that traces power solely to one class, group, or structure is likely to give us an incomplete understanding of the total system of domination in advanced societies (see also Manza, 1992: 292, 295).

The Role of the State

Giddens's dualistic view of power as both a human and a structural quality is roughly paralleled in his analysis of the state in advanced societies. In Giddens's opinion, Poulantzas and others are partly correct to represent the state as a structure or framework within which power is exercised by classes (or other interest groups) external to it. However, this does not negate the fact that these structured relations between positions within the state are also occupied by real people, by state leaders, bureaucrats, and lesser officials who retain special capacities or powers of their own (1981b: 218–220; see also Wright, 1994: 95–96).

Probably the most important sense in which the state is largely a structure for channelling the power of others is in its connections with the economy under capitalism. Giddens accepts the view most closely identified with Claus Offe, but roughly similar to that found in the work of other Marxists such as Poulantzas and Wright, that sees the state as dependent upon the activities of private capitalists for its revenue (Offe, 1974, 1984; Offe and Ronge, 1975). For this reason, the state structure to some extent is organized by state leaders to facilitate the economic goals of the bourgeoisie. In part, the state personnel are said to be caretakers or managers who, where possible, direct and augment the flow of capitalist economic power in order to aid and maintain both the bourgeoisie's surplus accumulation and the state's tax base. This is seen as a prime reason for the state's growth and its increased intrusion into economic affairs under advanced capitalism (1981b: 165, 211; 1990: 72; Giddens and Held, 1982: 192).

According to Giddens, the state in capitalist society also channels power in a somewhat different sense, by "insulating" political power from economic power so that these two means of control seem to the working class to be unrelated. Rather than rally the proletariat to achieve progressive economic change, perhaps even revolution, through political action, the prevailing political parties that have been instituted within most capitalist states keep these issues separate. For the most part, economic struggles are fought as labour–management disputes over wages and benefits in the industrial sphere, and pose no fundamental threat to the capitalist system itself. Thus, although workers have formal political equality with all other citizens in capitalist democracies, little thought is given to using politics to attain economic equality as well (1981b: 127–128; 1985: 322; 1990: 70).

Nevertheless, in acknowledging these ways in which the state is a structural conduit or framework for the power of external, especially capitalist, interests, Giddens asserts that the state is also a collection of social actors, of leaders and officials with considerable power in their own right. Here Giddens follows Weber to some degree, in that he stresses several bases for domination that are the state's particular preserve. First is the monopoly of organized violence inherent in the state's repressive (police and military) branches. Second is the capacity of these branches, as well as various legal and bureaucratic state organizations, to store information and strategic knowledge for surveillance purposes (1985: 2, 14–16; 1990: 56–60; 1994: 231). In addition, Giddens has recently stressed that states have formal control over the territory of their society, and also over the law-making apparatus (Giddens, 1998: 47–48, 53;

2000a: 122–123; see also 2002: 63–68; 2003: 13–17). Especially in the modern era, these means of domination and control give those who run the state special powers that are not subject to the will of the bourgeoisie or any other outside group. In contrast to the arguments of some other writers, Giddens asserts unequivocally that the governments of nation-states are "generically much more powerful" than even the giant multi-national corporations that have become so prominent on the international scene in recent decades (Giddens, 2000a: 122). Giddens sees the state's surveillance capabilities, control over the means of violence, and other powers as "independent influences" in shaping contemporary societies, no less significant for the process of social change than are capitalist class relations and class conflict (1985: 2). Moreover, such state powers are evident in virtually all modern societies, capitalist, socialist, or otherwise. Giddens finds it odd that only a few contemporary social theorists, either Marxist or non-Marxist, have paid much attention to these bases for state power. Clearly, there is ample evidence, throughout the twentieth century and beyond, of the use of such powers by totalitarian regimes on both the left and the right (1981b: 94, 175–177, 244; 1985: 297–300; 1994: 229–231; 2000b: 88, 96).

For Giddens, then, the state wields formidable weapons in the power structure of virtually all contemporary societies. At the same time, however, Giddens foresees several potential changes and developments in the current period that have affected the capacity of the state, as well as the capitalist class and other powerful factions, to shape the overall pattern of social inequality. These issues are central to Giddens's overall assessment of the prospects for inequality in future societies.

The Future of Social Inequality

Giddens's comments concerning the future of social inequality once again reveal a characteristic blend of Marxian and Weberian viewpoints. Like Weber, Giddens is highly critical of the various forms of state socialism that have arisen since Marx's time, and he continues to doubt that such systems can achieve the true transcendence of capitalism that they allegedly seek (1984: 32, 227; 2000b: 32; 2002: 66).

Nevertheless, like Marx, Giddens wishes there were a way of eliminating or transforming the exploitation, oppression, and other problems that are often associated with the capitalist system. This may explain why, at times, Giddens in the past supported the possibility of a democratic or "libertarian socialism" as a replacement for capitalism (1981b: 175). Although his more recent writings make it clear that his doubts about socialism remain unaltered, Giddens still believes that "some of the basic ideals associated with socialism … remain as persuasive as ever" and are "intrinsic to the good life" (1994: 247; 1998: 1).

To Giddens, the problem is that there are major shortcomings in both traditional capitalism and old-style socialism. First, consider capitalism. Giddens agrees that advanced capitalism has been successful on several counts, most notably the greater material affluence and political freedoms that it has provided compared with socialist and other nations. Unfortunately, modern capitalism has also fostered fundamental forms of injustice or disadvantage that have yet to be alleviated. The most basic difficulty with capitalism, of course, is that it is premised on the exploitation of one class by another. In addition, though, capitalism has typically been a party to ethnic, gender, and other important forms of exploitation. Moreover, while it is true that advanced capitalist societies have generally produced relatively more material affluence for people, including workers and other subordinate groups, serious problems of poverty still persist within these nations. Furthermore, it is arguable that at least part of the affluence in these societies has come at the expense of people in less developed countries elsewhere, through the global activity of modern capitalism and the international exploitation of nation-state by nation-state. Added to these considerations are growing problems of resource depletion and environmental degradation, which have been hastened by the globalization of capitalism and are pushing us closer to a worldwide ecological crisis (e.g., 1981a: 317–319; 1981b: 238–239, 250–251; 1985: 340–341; 1994: 100, 160, 246; 2000a: 132; 2000b: 39, 52; 2003: 33). To Giddens, neither the material benefits of capitalism nor the political rights and freedoms it offers have completely compensated for or eliminated these difficulties.

Giddens agrees with the Marxist view that one of the key problems with capitalism is that it is founded on a basic contradiction: capitalism is a system that is impossible without the social or collective production of surplus, and yet it also demands that a select few derive private and disproportionate benefit from that surplus. Thus, despite providing a measure of material security to workers, capitalism remains a highly unequal and problematic form of economic organization, one in which workers have little control over production and in which their labour is itself "commodified" and dehumanized (1985: 145, 311). Another important criticism that Giddens has posed is the rise of "productivism," a compulsive and destructive preoccupation with economic or material concerns that he says "stands close to capitalism" (1994: 247). In Giddens's view, such a way of life or style of thought has the negative effect of distracting people from real human "productivity," from the "personal growth" and "self-actualization" that comes through "living a happy life in harmony with others" (e.g., 1994: 166–169, 178–181, 195). While a certain level of economic growth and material well-being is necessary for human happiness, a major failing of contemporary capitalism is that it has promoted an excessive focus on such matters.

Although these weaknesses do not mean that capitalism is in imminent danger of collapse, Giddens does suggest that capitalist societies are subject to chronic strains and pressures. In recent years, the state has played a role in containing some of these tensions, by trying to maintain social conditions that are acceptable to the general population without at the same time undermining bourgeois privilege. However, Giddens's view is that the "welfare state" programs of contemporary capitalist governments, involving the provision of various social services to much of the population, have met with only mixed success at best. Too often such programs are simply "reactive" attempts at "damage control." They are not "positive" or "generative" solutions, and deal more with the symptoms than with the actual causes of poverty, underprivilege, and so on (e.g., 1994: 4, 151–157). In particular, welfare policies in most nation-states have not been geared to giving poor people the means to break out of the poverty trap resulting from long-term dependency on government support. Giddens advocates much stronger state efforts to enhance people's "human capital" by providing employable skills and job retraining, for example. In this way, provided that government also maintains a sense of fiscal responsibility and lives within its means, the state can play a more effective role in giving a genuine "hand-up," rather than a barely adequate "hand-out" to those in need (e.g., Giddens, 2000a: 73,106; 2003: 2–6).

Because of the problems that Giddens identifies in contemporary capitalism, we might expect some renewed prospect for socialist revolution in the future. As noted earlier, however, Giddens sees such an outcome as unlikely. On the contrary, he contends that the forms of socialism we have seen throughout the twentieth century are "surely fading away" (1990: 164; 1994: 8). He is convinced that "at least as a system of economic management, socialism is no more" (1998: 3, 43; see also 2000a: 28, 56, 164; 2000b: 32). One of the major difficulties with socialism has been its clear and chronic failure to fulfill its promises. Giddens has described the various examples of state socialism to date as no more than "clumsy, abortive prototypes" that cannot transcend the capitalist system until they come to grips with their own contradictions (1981a: 319–320). Socialism's central contradiction has always been that it seeks mass equality and participation in social policies and decisions, and yet requires a centralized system of production and administration that focuses power in the state. As we have already noted, the concentration of state power, especially through control of surveillance and repression, is a serious threat to equality and freedom that is at least as likely under socialism as under capitalism. Similarly, the bureaucratic domination, sexism, racism, and colonialism common to many capitalist systems have also been prevalent in state-socialist regimes (1981b: 242–243, 248–252). Thus, whatever direction humanity may take in the future, for Giddens the preferred course "would not be socialism," or at least not the flawed versions of socialism that have arisen so far (1994: 159).

One crucial lesson here for Marxists is that the abolition of class exploitation through socialist revolution is not enough by itself to bring about a just society, since various other problems and bases for exploitation still persist. The inability of many Marxists to perceive this fact stems in part from their singular concern with class and their consequent failure to note that class relations, though important, are but one manifestation of power differences.

Giddens has also suggested that the role played by class, both as a basis for group solidarity or identity and as a force for social change, will be altered, and perhaps weakened, in the future. In part, Giddens attributes this trend to continual changes that he believes are occurring in both the occupational structure and people's employment histories. These include the shifting back and forth between blue-collar and white-collar jobs by some workers and the regular movement in and out of newly emerging types of occupations, most of which are quite different from those performed by previous generations. Many of these developments stem from the expansion of jobs in the new "knowledge-based" communications and information technology areas of the burgeoning global economy. Giddens suggests that those who adapt to these new and largely irreversible circumstances probably will experience class as less of a "lifetime experience" than their parents did, while those who do not accommodate to the changing economic realities will face the prospect of diminishing employability and the decline of paid work as a central element of their existence (1994: 142–145; 2000a: 65–69). Such changes could lead to a movement away from class position as the core focus of human existence and identity. This argument for the possible reduced salience of class is also consistent with Giddens's assertion that if there is to be an end to exploitation and oppression in the future, "there is no single agent, group, or movement that, as Marx's proletariat was supposed to do, can carry the hopes of humanity" (1994: 21).

Given the serious flaws that Giddens perceives in both capitalism and socialism, we might wonder if any system can be devised that will solve the fundamental problems that now loom in the world. Without some workable alternative to the existing approaches, there is little reason for optimism about the future. Indeed, at times Giddens has expressed considerable pessimism in this regard, most notably his worries about the possibility of total nuclear destruction (1981b: 196–198, 250; 1985: 334–338). Giddens has delineated four major concerns, or "global bads," that currently confront humanity (1994: 100, 253). These include the polarization of rich and poor nations or peoples, threats to the world's environment or ecology, the denial of democratic rights in authoritarian systems, and the continuing problems of violence and the risk of large-scale war (see also 2000b: 33–34; 2002: 70–72).

However, while Giddens believes that all of these problems could yet lead to our complete demise, his more recent view has been that humanity may actually have more hope than in the past, that today there is at least some "good cause for optimism" (1994: 21; 2000a: 132; 2000b: 99–100). Ironically, Giddens traces much of this optimism to developments that previously were sources of concern and anxiety. For example, economic globalization has made some nations very rich and has given great power to a relatively small number of transnational corporations. At the same time, though, this process has also "decentred" economic control to some degree, with more international business competition and a weakening of some of the monopolies that once flourished within national economies (1994: 87–88). Similarly, Giddens has suggested that globalization has increasingly undermined the "collaborative connections between state and capital" that in the past contributed to the rise of monopolies and the concentration of both economic and political power (1994: 89). As a result, although individual national governments and the capitalist classes of specific countries are still tremendously powerful, virtually none of them can operate with the impunity that they may have enjoyed in the past (e.g., 1994: 89, 142–143; see also 2000a: 143–150; 2000b: 96–98).

Another important illustration of recent changes concerns the threat of nuclear war, which Giddens has usually tied to his discussion of the monopoly over the use of violence within modern nation-states. Here, as well, Giddens suggests that some of the risks have receded, paradoxically because of the global spread of the nuclear threat itself. In other words, the further development, greater destructive potential, and wider availability of nuclear weapons now make their use still possible but probably less likely, given the devastating consequences that would result, not only for the combatants but for virtually everyone else in the world (1994: 232–235; 1998: 139–140). On the other hand, Giddens is aware that, since the terrible events of September 11, 2001, the possibility of global terror, through nuclear weapons or other lethal means, is a completely new threat in some ways and a new cause for pessimism; even so, this threat too is something that affects everyone on the planet, and it is a problem that also will require a global, international, and multilateral response if it is to be resolved (Giddens, 2002: 71–72; 2003: 29–31).

Additional forms of violence and confrontation remain, of course, including numerous ethnic, nationalist, and religious conflicts between or within countries (1994: 233, 242–245). Equally worrisome is the serious problem of domestic assault and violence against women in much of the world (1994: 236–239). In these instances, too, however, Giddens holds some hope for the future, again in part because of certain globalizing tendencies. In particular, he points to advances in modern communication systems, which have multiplied the capacity of people to transmit ideas, knowledge, and information. Such developments have helped move us closer to a "post-traditional social order," marked by greater "reflexivity" and "cosmopolitanism" than in the past (e.g., 1994: 4–7, 42, 80–83; 2000a: 157–158; 2000b: 22–23, 90, 96; 2002: 71). The almost instantaneous nature of communication, which Giddens believes is a central element in the globalization process, promotes an expanded awareness among all of us that different societies inevitably and increasingly share the whole planet, and that there are other ways of thinking and acting that diverge from our own. This growing awareness could even make possible the eventual establishment and acceptance of "universal" values in the world, including the sanctity of human life, the right of everyone to happiness and self-fulfillment, and the commitment to genuine democracy (1994: 20, 253; 2000b: 68; 2003: 4). It also means that all the crucial problems to be solved—not only war and violence but also poverty, authoritarianism, bigotry, ecological disaster, and terrorism, among others—must in a very real sense be understood as everyone's problems.

Of course, awareness of the world's problems is by itself insufficient to bring about their resolution. Some prescription or plan for solving these global difficulties is also required. It is here that Giddens has proposed a "third way," an approach that goes beyond the flawed versions of both capitalism and socialism that we have witnessed over the past century or more. Some have suggested that Giddens's third way is a largely pro-business or pro-capitalist scheme. To Giddens, however, such charges are true only in the sense that he now agrees with those who believe, first, that there is no longer a viable socialist alternative to some form of capitalist market economy and, second, that there is no going back from the trend toward the globalization of this economic system. Nevertheless, Giddens contends that he is actually on the "modernizing left," someone who believes that the best course is to develop a "modernizing social democracy" based on "neo-progressivism" (Giddens, 2000a: 27; 2003: 6; Giddens and Diamond, 2005). Such a society should be able to harness the new world economy so as to maximize its positive consequences, and redirect its productive capacity for the material and social betterment of as much of the world's population as possible (Giddens, 2000a: 22, 124–125, 164–165; 2002: 72–75).

Among other key requirements, of course, are effective agents of change, groups of people who have the capacity and the willingness to apply this third way, so as to resolve the important difficulties that the world faces. On this question, Giddens's key proposal is an international impetus toward truly effective democracy. What may be needed most is the "democratizing of democracy," by which Giddens means the inclusion of many more agents, as well as new types of agents, in the processes by which decisions are made and policies implemented in societies (Giddens, 1998: 70–78; 2000a: 61–62, 139–140; 2000b: 93–97). In other words, we can no longer get by with the old arrangements, in which party leaders, whether from the left or the right of the political spectrum, as well as big owners (and occasionally big unions) from the economic sphere, were able to dominate or control much of the democratic agenda. Nor can we accept that the major decisions should be left to so-called experts in these and other areas. Instead, the solution to global difficulties is to engage more of the global population itself, most notably in the form of citizens' groups and other organizations that exist outside of the formal economic and political realms, in what is typically referred to as "civil society." For Giddens, striking a balance of countervailing powers involving "government, the economy, and civil society" is the most effective way to achieve the betterment of all people and "a healthy global order" (Giddens, 2000a: 123, 51–52, 144; 2002: 63–68). His examples of agents from civil society comprise a mix of different organizations and collectivities operating at the local, national, and international levels. They include groups associated with the so-called "new" social movements centred on issues of gender, the environment, world peace, and so on. All of these agents, as well as "self-help" groups and other organizations, must be part of a fundamental impetus toward constructive social change and an open, participatory, or "dialogic" style of democracy. In this way, Giddens believes that we may yet succeed in achieving "the social revolutions of our time" (1994: 15–17, 78–79, 249–250; 1998: 47, 53; 2000a: 123, 140, 157; 2002: 63–68; 2003: 21, 28).

Giddens's observations and prescriptions for the future share some interesting similarities with ideas we have seen before, from writers such as Marx and especially Durkheim. In particular, Giddens echoes Durkheim in suggesting that many contemporary problems are essentially problems of "solidarity," resulting from "egoism," "demoralization," and a "crisis of moral meaning" (e.g., 1994: 13, 124, 195, 247). His preferences for the future are reminiscent, as well, of Durkheim's normal division of labour, and to some extent Marx's true communism, in that Giddens envisions a society where people are bound together by "mutual responsibility," "reciprocal obligations," and a real sense of "trust" in others (e.g., 1994: 12, 126, 195). There are clear similarities with Durkheim, as well, in Giddens's hope for a new "institutional" form of individualism, which "is not the same as egoism" and which does not pose a threat to "social solidarity"; individualism in this sense means that there should be "no rights without responsibilities," that people must balance their own sense of personal entitlements with a fuller appreciation of their commitments to others and to the collective good (1998: 36–37, 65; 2000a: 52; 2002: 16; 2003: 3–4).

Giddens also parallels both Durkheim and Marx in playing down the importance of economic equality as the crucial ingredient for a good society. Giddens argues instead for a "post-scarcity society," in which the achievement of equality "is based less on a rigorous sharing of material things than an indifference to them" (1994: 191). His more recent work, however, does make some allowance for the continuing existence of economic inequality in future societies. Although he strongly advocates a significant redistribution of income and wealth from the rich to the poor, including Marx's and Durkheim's proposal of using taxation to eliminate the inheritance of economic privilege across generations, Giddens nevertheless explicitly argues that some inequality, provided it is based on equality of opportunity, is acceptable and even desirable, because "the possibility of becoming wealthy ... may motivate exceptional talent" (Giddens, 2000a: 96–97, 86–89; 2002: 16–17, 25; 2003: 3–4; see also Giddens and Diamond, 2005).

One potential criticism of Giddens's discussion of the future is that it may have become too optimistic, or an exercise in wishful thinking. Giddens actually has partly conceded this point, acknowledging that his viewpoint, which he has at times referred to as "utopian realism," is currently "more possibility than actuality" (1994: 101, 124). He has conceded, as well, that the risks of serious global crises persist and that because of our growing interdependence in the world, "when things go wrong, they can go very wrong" (Giddens, 2000a: 162; 2003: 28–30). Giddens disagrees, however, with those who claim that his ideas are far too utopian to be realistic (1994: 248). He also leaves his critics with a profound and telling question (1990: 165): if human beings are somehow to deal with global problems and avoid self-destruction, "what other alternative is there?"

Summary Observations

The purpose of this segment of the chapter has been to outline and assess Giddens's conception of social inequality, with special reference to his structuration approach to class, power, and other related issues. Of the recent perspectives we have examined, Giddens's approach in several respects may offer the most comprehensive and inclusive strategy for thinking about social systems and the inequalities that inhere within them. In addition to retaining Marx's classical concern with class and Weber's pluralist revisions and modifications of Marx, Giddens attempts to incorporate what he believes is the central strength of structural functionalism: the delineation of major institutions and their attendant concrete structures in advanced societies. The economic, political, religious, and other structures that are identified in this way can all, in varying degrees, be conceived as systems of power, in which patterned relations of domination are established and reproduced over time and space, based on differential control of material ("allocative") and human ("authoritative") resources.

Although there are also clear differences in approach, Giddens's recognition of these distinct structures of domination roughly parallels the separation of economic, political, and ideological apparatuses suggested by Poulantzas, Wright, and other recent Marxists. The advantage Giddens has over these conceptions stems mainly from his more general concern with analyzing all power relations arising within these structures, as opposed to focusing almost exclusively on class relations (see also Mann, 1986, 1993; Runciman, 1989).

Giddens's stress on the universal role of power in generating all forms of inequality has some affinities with Dahrendorf, closer connections with Lenski, and even greater similarities with Parkin's social-closure conception. There is a pleasing clarity in Parkin's viewpoint that contrasts with Giddens's rather complex formulation and with Giddens's propensity, as Parkin notes, "to drive conceptual wedges between empirically inseparable things" (Parkin, 1980: 892; see also Giddens, 1980). Giddens's attempt to differentiate "locales" from "places" in social space illustrates this occasional tendency to obscure, rather than clarify, his discussion with rather subtle or minor distinctions (1985: 12). Another example is his insistence that there is an important conceptual difference between the idea of "structure," on the one hand, and the idea of "structures," on the other (1984: 18; see Turner, 1986: 972; Sewell, 1992; Lemert, 1995). Nevertheless, some of the complicated aspects of Giddens's analysis can also be seen as evidence of its completeness. It would appear that Giddens, more than Parkin, offers us a perspective on inequality that acknowledges the full range of social situations in which power is exercised and inequality established. Giddens also combines Parkin's astute awareness of power as a property of real people with the neo-Marxist (and structural-functionalist) sense of structural arrangements as repositories and channels for power that exist, in a way, apart from the persons that staff them.

For these reasons, then, it could be argued that in certain ways Giddens's overall approach provides the most fully developed guide for thinking about and analyzing social inequality that currently exists in social theory. Inevitably, of course, there are specific points raised by Giddens, and by the other theorists for that matter, that remain open to debate and that are unlikely to be reconciled to the satisfaction of every theoretical camp. Nevertheless, it should be evident by now that the writers we have examined are concerned with many of the same issues and, to different degrees, have all contributed to a more complete appreciation of what social inequality is and how it should be conceived.

Before concluding our analysis, it is important to review some other theoretical developments in contemporary approaches to the study of social inequality. In particular, it is useful to assess several of the major insights that have emerged from recent discussions of two specific forms of social inequality: gender and race/ethnicity. In the next two sections of this chapter, we examine some of the significant advances in these two more focused areas and suggest key ways in which these discussions can also be seen as contributing to our general understanding of how social inequality may best be analyzed and portrayed.

At the conclusion of this chapter, we shall also briefly discuss work reflecting the growing interest that recent theorists and researchers have shown in the global nature of social inequality in the contemporary period.

LINKING GENERAL THEORIES TO GENDER AND RACE

Theories of Gender Inequality

It is probable that, in the last two or more decades, no single topic in the study of inequality has received more attention than the analysis of gender. The numerous perspectives on gender that have emerged in the contemporary literature span a wide range of sociological thought, including everything from Marxism to neofunctionalism (see, e.g., Wallace, 1989; England, 1993; Saunders, 1999; Walby, 1999; McMullin, 2004: ch. 3; Siltanen, 2004; Nelson, 2005). Still, the large majority of the most influential gender analyses clearly fall toward the left end of our left–right continuum, since they generally portray gender inequality as a product of collective struggle and power inequities between the sexes.

Gender and the Classical Theorists

The substantial growth in theoretical discussions of gender in recent years has helped to compensate for the neglect of this topic in many early analyses. Several observers have suggested, in particular, that the classical theories of society developed by Marx, Weber, and Durkheim are among those that seem to be "gender-blind," in the sense that they usually disregard or underestimate the importance of this question for understanding the workings of society in general and social inequality in particular (e.g., Sydie, 1987: 1–3; Bologh, 1987: 145; Sayer, 1991: 32, 108; Collins, 1994: 119; Adams and Sydie, 2001: 114). The tendency of the classical writers to overlook or give only limited attention to gender can be traced primarily to the times in which they lived and worked. All three men were essentially nineteenth-century thinkers, with ideas and interests shaped in large part by the male-dominated intellectual climate of that era. Of course, it is also a fact that, during that period, women were excluded from all but a few positions of influence or prominence, both in academia and elsewhere. Thus, in a sense, the absence of detailed discussions of gender issues in classical sociological theory is itself an accurate reflection of the inequalities that women faced at that time.

Researchers have pointed out that a number of feminist thinkers and prototypical gender theorists, including Charlotte Perkins Gilman, Harriet Martineau, Beatrice Potter Webb, and others, lived and wrote during roughly the same time period as Marx, Weber, and Durkheim (see, e.g., Lemert, 1995: 117, 202; Adams and Sydie, 2001: 267–290; Kivisto, 2003: chaps. 25–29; Ritzer and Goodman, 2003: 271–303). The failure of the early theorists to pay heed to these pioneering efforts in gender analysis provides further evidence of the male-oriented nature of social science research and scholarship that prevailed in that era.

However, while it is true that Marx, Weber, and Durkheim do not provide complete or extended analyses of gender, it would be incorrect to argue that they are heedless of women's subordination or disadvantage in society. To varying degrees, all three classical figures show an awareness of such issues, but each thinker discusses gender, not as a separate question, but mainly in terms of his own principal theoretical preoccupations.

As might be anticipated, Marx generally blames gender inequality on the injustices of the capitalist class structure. For him, class inequality is the central component in an economic system that not only exploits male workers but also treats "women as mere instruments of production" (Marx and Engels, 1848: 89; Marx, 1863: 200). Engels, who wrote more extensively than Marx on this subject, argues similarly that gender stratification is ultimately determined by economic forces and capitalist class relations. Engels elaborates on the Marxian argument by suggesting that, in addition to being exploited by the bourgeoisie when working outside the home, women are doubly exploited in their roles as domestic labourers (Engels, 1884). That is, females are burdened with most household duties, usually in exchange for economic support by and subordination to their male spouses. This arrangement also benefits the capitalist class, because women's unpaid labour—feeding, clothing, comforting, and caring for their husbands and children—in effect makes it possible for male workers to perform their daily toil for the bourgeoisie (Engels, 1884; see also Sydie, 1987: 89–101; Collins, 1988b: 165–168; 1994: 78–81; Adams and Sydie, 2001: 136–138). According to classical Marxism, these and other gender inequities are byproducts of capitalism and will disappear in the transformation to socialism.

In Durkheim's work, as well, the discussion of gender issues is usually connected to his more central theoretical concerns, most notably the division of labour. It will be recalled that Durkheim foresees a trend toward a normal division of labour in future societies, with people enjoying a high level of individual freedom and equal opportunity to find their appropriate positions in life. However, Durkheim's view of the future has some important restrictions, for his ideas of equal opportunity and freedom are focused primarily on the sphere of paid employment and apply almost exclusively to men (e.g., Sydie, 1987: 14–24; 1994:

122–126; Collins, 1988a: 119; Lehmann, 1994; 1995a, 1995b). Durkheim says little about female participation in the labour force itself, but he appears to prefer a traditional "sexual division of labour," in which women's activities are concentrated in the family or domestic setting rather than the occupational structure (Durkheim, 1893: 56; 1897: 385; see Sydie, 1987: 24; 1994: 125; Lehmann, 1995a: 919; Adams and Sydie, 2001: 108–114).

A somewhat similar view of sex-role differentiation has been sustained by subsequent structural functionalists. Parsons, for example, envisions a situation in which men engage largely in "instrumental" or occupational roles outside the home, while women predominate in the "expressive," caregiving roles of the domestic sphere (e.g., Parsons, 1954). It has been pointed out that this arrangement need not mean that females are subordinated to males, but rather that men and women perform different, complementary tasks (see Johnson, 1993; Saunders, 1999). It should be noted, as well, that neither Durkheim nor Parsons precludes the participation of women in paid employment. Nevertheless, both Durkheim's writings and those of later structural-functionalist theorists imply that it is generally better for the well-being of the family and society if women's interests and aspirations outside the home are subordinated to those of men (but see Tiryakian, 1995).

Finally, there is Weber, who, compared with Marx and Durkheim, probably provides the most complete and systematic treatment of gender inequality (see Collins, 1986a: 269–270; Adams and Sydie, 2001: 189–192). In a manner consistent with his general theoretical emphases, Weber sees gender inequality in "political" or power terms, arising especially from relations of domination within the family or household setting. Weber argues that, beginning with the earliest forms of society, women usually find themselves in a subordinate situation relative to men. This subordination stems from the greater coercive powers of males, whose superior physical strength and attendant military and protective roles typically put them in a position of dominance. Over time, the coercive aspects of male domination are likely to become less overt, and also to be legitimized within the family and most other societal institutions. Still, the actual or potential use of physical force tends to underlie gender relations and bolsters both the economic and the sexual domination of women, as well (Weber, 1922: 359, 364; 1923: 48; for discussion, see Collins, 1986a: 271–285; 1988b: 168–173; 1994: 79, 120). In addition, as was noted in Chapter 3, the greater power accorded males in such a relationship can also establish gender as a basis for what Weber calls a status group, since men are able to exclude women from certain privileges and prestige simply because of their being male.

At numerous points in his discussions of gender, Weber draws on the notion of *patriarchy* (e.g., Weber, 1922: 372; 1923: 46). Patriarchy has since become one of the most widely used ideas in contemporary gender theory, and at times has taken on rather different meanings. For example, some reserve the term to refer to male-dominated family structures, while others use it as an ideological or cultural concept, denoting sets of values or beliefs that support male authority (for discussion, see Connell, 1987; Fox, 1988; Acker, 1989; Walby, 1989, 1990, 1997; Mackie, 1991; Lips, 1991; Shelton and Agger, 1993; Adams and Sydie, 2001: 182–183). More generally, though, patriarchy can refer to any structured pattern of relations within society in which males regularly have power over and therefore dominate females (e.g., Walby, 1989: 214; 1990: 20; Hurst, 1995: 82–83; McMullin, 2004: 45–49; Siltanen, 2004: 222–224; Nelson, 2005: 87–88). Although Weber's analysis of gender is incomplete, he is notable nonetheless for being one of the first theorists to indicate that patriarchy, or male domination in general, lies at the root of gender inequality. Weber's work also suggests that such domination has become a pervasive feature of most social structures and also manifests itself in a multiplicity of ways—physically, economically, sexually, and so on. Here Weber again reveals his fundamental view that the causes of social phenomena must be understood in pluralist terms, and that all forms of inequality arise mainly from power differences among groups.

In summary, the classical theories of Marx, Durkheim, and Weber reveal some obvious deficiencies and omissions in their approaches to gender issues. At the same time, however, these early perspectives contain insights that, with some amplification, can contribute significantly to our understanding of the bases for gender inequality in contemporary societies. Three points in particular stand out. First is the realization, drawn mainly from Marxian theory, that class relations intersect with and can reinforce women's exploitation or disadvantage. Second is the acknowledgment, which is evident to some extent in all three classical approaches, that a major reason behind female subordination is the manner in which labour is divided, especially the distribution of household or domestic work. Finally, there is the recognition, found primarily in Weber's writings, that gender inequality exists in a plurality of social structures or institutions and is traceable primarily to differences in power between men and women. These three basic insights contribute significantly to an enhanced understanding of gender inequality in society. It can also be argued that, in one way or another, the best recent perspectives on gender inequality have devoted much of their efforts to developing precisely the same three ideas. Let us briefly consider how each of these points has been elaborated and extended in the current literature.

Connecting Gender with Class, Domestic Labour, and Power

Most leading theorists, whether they are Marxists or non-Marxists, feminists or generalists, now readily accept the view that gender inequality is usually exacerbated by, and often closely tied to, class inequality. We have already seen evidence of this idea in the work of Wright and Giddens, but many other contemporary writers emphasize the same gender–class connection. Several often include in their analyses the linkages to race or ethnicity, as well. In fact, there is a growing awareness that all of these forms of inequality interact and overlap, in what some have called a "matrix of domination" (Andersen and Collins, 1995: xi; see also Armstrong and Armstrong, 1990: 68–77; Stasiulis, 1990; Chafetz, 1990: 196–198; 1999; Balibar and Wallerstein, 1991: 9, 49; Agger, 1993: 2, 77–82; Ward, 1993: 43; Rothman, 1993: 2–6; Clement and Myles, 1994: 123–28; Jackman, 1994; Hurst, 1995: 100–114; Walby, 1997: 6; Siltanen, 2004: 225–226; McMullin, 2004: 54–55).

One sign of the increased awareness of the association between class and gender inequality is the large number of important debates that have recently emerged over how they are interconnected. An example is the controversy about the suggestion that gender may be a relatively more important factor than class in the overall structure of inequality, since female subordination predates the capitalist class system and has existed in virtually all known forms of society (see, e.g., Hartmann, 1981; Connell, 1987; Wilson, 1993; Shelton and Agger, 1993; Hurst, 1995; Walby, 1997; recall also Giddens, 1981b; Wright, 1994). A second illustration concerns the dispute over Goldthorpe's claim that all members of a family occupy the same class location, which he says is normally determined by the occupation of the male household head (e.g., Goldthorpe, 1983; 1984; Goldthorpe and Payne, 1986). Some observers have questioned this claim, arguing that it ignores the potentially different class locations of female spouses, who increasingly hold jobs of their own outside the home (e.g., Stanworth, 1984; Walby, 1986, 1997; Wright, 1997; but see Baxter, 1994). Both of these examples demonstrate the often contentious and complex issues that can arise when we consider the intersection of gender and class inequality in contemporary societies. These debates and others, many of which remain unresolved, continue to stimulate and enliven the theoretical discussions in the general field of social inequality.

The second key insight that is now a major theme in recent gender theory is the importance attached to the division of both domestic and paid labour. The early Marxian view that female household labour indirectly serves capitalism has been accepted and expanded by many contemporary writers, especially Marxist and Marxist-feminist theorists. Although some analysts disagree on finer points and more specific aspects of the argument, most concur with the idea that women's domestic toil helps to "reproduce" labour power for the bourgeoisie, by maintaining a household environment that allows males to work for a living wage in the employ of the capitalist class (see, e.g., Oakley, 1974; Acker, 1989;

Murgatroyd, 1989; Armstrong and Armstrong, 1990; Shelton and Agger, 1993; Agger, 1993; Siltanen, 2004). In addition, these theorists and others have shown that domestic activities are just part of what is usually a "double day" for females, since women in most societies are increasingly involved in the paid labour force outside the home. There are at least two implications for gender inequality that arise from this double burden on females: first, women are typically unable to compete on an equal basis with men for the best-paid jobs and most powerful positions in society since most women must devote so much of their energy to household labour; second, the types of work that women perform in the paid division of labour have a consistent tendency to segregate females, not only in lower-paying and less influential positions but also in jobs involving duties that parallel their traditional roles in the home, as caregivers, service providers, and child supervisors (see, e.g., Luxton, 1980; Armstrong and Armstrong, 1984; Reskin and Hartmann, 1986; Reskin and Roos, 1990; England, 1992; Reskin and Padavic, 1994; Siltanen, Jarman, and Blackburn, 1995; Walby, 1997; Siltanen, 2004). These features of the gender-based division of labour are said by many leading theorists and researchers to be among the most important reasons behind gender inequality in society (Chafetz, 1984, 1990; see also Huber and Spitze, 1983; Blumberg, 1984, 2004; Crompton and Mann, 1986; Milkman, 1987; Lips, 1991; Clement and Myles, 1994; Siltanen, 1994, 2004; Blumberg et al., 1995).

The third major development in recent perspectives on gender is the central significance that is now attached to power inequities. We have noted that Weber was one of the first theorists to emphasize power, and the related notion of patriarchy, in his discussions of gender inequality. Now, however, power or domination has become the core element in many of the best theoretical discussions of gender to emerge in the current literature. Some of these discussions have extended earlier analyses of power in important ways. Janet Chafetz, for example, has demonstrated how power differences between men and women can develop and operate at various levels, in large-scale "macro-structures" like the economic system or the state, in intermediate-level "mezo-structures" such as the workplace or community organizations, and in more intimate "micro-structures," most notably the family setting and romantic couples (Chafetz, 1990: 15, 21–23, 1999; see also Blumberg, 1984, 1991a, 2004; Blalock, 1989; Huber, 1990, 1991; Dunn, Almquist, and Chafetz, 1993; Andersen and Collins, 1995: 191–200). Echoing Weber to some extent, Chafetz also notes that power differences between genders may not always be formal and "coercive" in nature, but at times can arise from more informal and "voluntaristic" processes, as when women themselves adopt beliefs and make life choices that "inadvertently contribute to their own disadvantage" (Chafetz, 1990: 19, 64–81; see also Bem, 1993).

Another significant contribution to the discussion of gender and power can be found in the work of Sylvia Walby. Although her arguments are couched primarily in terms of patriarchy, rather than power or domination per se, Walby's analysis parallels Chafetz in suggesting that there are multiple power structures in society operating at different levels of abstraction. She identifies six main "patriarchal structures," or what might be termed structures of gender domination, within which the overall pattern of gender inequality is defined and played out in social practice (Walby, 1989: 220–229; 1990: 19–24). These include: the structure of household labour; the structure of paid labour outside the home; the structure of the state or government; established practices within sexual relations; established "cultural" practices that promote patriarchal ideologies and beliefs, as in religion, the media, and the education system; and the recurring incidence of male violence against women, which in itself can give rise to regularized patterns of female subordination in social settings. Walby's delineation of these six structures matches Chafetz's, and to some extent Weber's, view that gender inequality is shaped by gender differences in power that exist in various social locations, both public and private, both macroscopic, microscopic, and in between. Walby's work here also hints at the notion that several different types of power—for example, economic power differences existing inside and outside the home, political power differences within the state, and ideological or cultural power differences embedded in religious and educational institutions—are involved in the overall structuring of gender inequality (see also Walby, 2002, 2005).

In fact, using an approach broadly similar to that of theorists like Poulantzas, Wright, or Giddens, many recent gender analysts now stress the importance of understanding the somewhat different roles played by these economic, political, and ideological/cultural types of power relations (see, e.g., Scott, 1986: 154; Chafetz, 1990: 31; Dunn et al., 1993: 72; Andersen and Collins, 1995: 191; see also Connell, 1987: 34–35; Mackie, 1991: 198; Lips, 1991: 109). Of course, within this range of theorists, there is considerable debate over which sources of power are more significant or fundamental in contributing to male domination. Many theorists, including radical or Marxist feminists but also various non-Marxists, stress the importance of economic power differences, especially regarding control over material production and the distribution of economic surplus. Despite differences in emphasis and tone in these accounts, their common element is the acknowledged significance of economic power as a major factor contributing to female subordination and the overall pattern of gender inequality (e.g., Blumberg, 1978, 1984, 1991b, 2004; Chafetz, 1984, 1990, 1997; Blumberg et al., 1995; see also Friedan, 1963; Millet, 1969; Bernard, 1971; Eisenstein, 1979; Huber and Spitze, 1983; Crompton and Mann,

1986; for reviews, see Richardson, 1988; Ritzer, 1988; Collins, 1988b; Mackie, 1991; Lips, 1991; Hurst, 1995).

In addition to economic power, however, many prominent writers in this field conceive of political power as another major force accounting for the overall domination of men over women in social structures. Some suggest, in fact, that political (including coercive or military) power has at times played the most crucial role in establishing male dominance historically (Collins, 1975: 225–259; Collins, 1988b: 168–173; see also Millett, 1969; Chafetz, 1990: 86–88; Lips, 1991). Others acknowledge the importance of political or coercive mechanisms but still see economic power as more significant (Blumberg, 1984: 41, 49, 74–75).

Finally, along with economic and political power, many theories that address the problem of gender inequality place considerable weight on ideological domination, on the power that ideas, beliefs, and cultural values can have in establishing the control of one group or faction by another in social settings. Most of these perspectives contend that the teaching and dissemination through society's ideological structures of what roles males and females should occupy can be of major consequence in determining the pattern and extent of gender inequality (see Blumberg, 1984: 40–41, 45; Ritzer, 1988: 306–307; Richardson, 1988: 156–158; Blalock, 1989; Chafetz, 1990: 35–37; Walby, 1990: 21, 104–108). Some even contend that ideological factors are, if anything, more important than economic or political considerations (e.g., Sanday, 1981; Milkman, 1987: 153). Theorists and students of patriarchy seem especially likely to mention the part played by ideological power, and parallel many of the classical and contemporary theorists we have reviewed, in including this element in their explanations of gender inequality.

We can see, then, that there are some important affinities and correspondences between the approaches used by leading recent theorists of gender inequality and those that we have found in the general classical and contemporary perspectives. This is not to suggest that theories focusing specifically on gender issues can simply be subsumed under the more general approaches. For a variety of reasons, the explanations for gender inequality are likely to differ in certain ways from those that deal with class, race, or other forms of inequality. For example, the division of domestic labour probably affects gender inequality more than any other type of inequality in most societies. There are also likely to be variations in the relative impact of economic, political, and ideological power for each type of inequality. And, of course, explanations for gender inequality will themselves vary because of differences in how power is distributed to men and women across a range of historical periods, geographic areas, organizational levels, and institutional settings. However, a principal claim here is that gender parallels the other key bases for inequality in that, first, it can be understood using much the same conceptual vocabulary and, second, it arises largely because of differences in access to the three major sources of power: economic, political, and ideological.

Theories of Racial or Ethnic Inequality

As with the analysis of gender inequality, the study of race and ethnic relations has witnessed a significant resurgence in recent years. Much of this new research has come from writers working in the same broadly Marxist or Weberian traditions that have dominated the field of social inequality for the past several decades. The increased attention given to race and ethnicity in the current period is probably less marked than that which has been devoted to gender. However, both of these growing literatures are notable for expanding our understanding of these topics, and also for filling in some of the conceptual gaps left by classical theory.

Race, Ethnicity, and Classical Theory

In their relatively brief treatments of the subject, Marx, Durkheim, and Weber approach race and ethnicity in a way that is consistent with their general perspectives. Marx, as we would expect, is primarily concerned with how ethnic or racial inequality relates to his overall analysis of the capitalist class structure. His view is that racism and ethnic oppression are largely vestiges of the pre-capitalist era. Such problems may continue in capitalism for a time but will gradually disappear in the march toward socialism. Thus, by the time of the revolution, Marx believes that the bourgeoisie will be faced with an international proletarian movement, led by workers who "have no country," who are motivated by the class struggle and not by ethnic conflicts or "national differences and antagonisms" (Marx and Engels, 1848: 90, 64; see also Sayer, 1991: 49, 53, 81; Adams and Sydie, 2001: 138–139).

Durkheim's treatment of race and ethnicity is also subsumed under his more pressing theoretical interests, especially the division of labour (e.g., Fenton, 1980, 1984; Lehmann, 1995b). Not unlike Marx, Durkheim sees racial or ethnic inequalities mainly as symptoms of an unfair economic system, the so-called "forced" division of labour. He suggests that injustices based on race or ethnicity, just like those based on class, should eventually be eliminated as society advances toward the "normal" division of labour. Under this new system, ideas such as race and ethnicity may actually lose their meaning and disappear altogether, since there will be equal opportunity for all, and people will be judged only on their individual merits and abilities (Durkheim, 1893: 304; 1897: 85).

Finally, as for Weber, he lists race or ethnicity among the numerous bases for status group formation in society (Weber, 1922: 932–936). In a related way, as was noted in Chapter 3, he also discusses the activities of competing or conflicting "interest groups," who often use an "externally-identifiable characteristic" like race as a pretext for excluding people from important rights and privileges (1922: 342). Weber's assessment here is coupled as ever with his consistent emphasis on the role of power relations in explaining virtually all forms of inequality in society.

Race and Ethnicity: The Role of Class, Conflict, and Power

Much of the recent literature on racial and ethnic inequality has concentrated on extending some of the ideas left undeveloped in the classical perspectives. Thus, some Marxist scholars, while agreeing with Marx that racism should be dispelled as we progress toward the end of the capitalist system, have suggested nonetheless that the effects of racial and ethnic strife can be more persistent under capitalism than Marx realized. According to this argument, the continuation of racial and ethnic prejudice is often actively or tacitly encouraged by the bourgeoisie. As occurred in the old South Africa, for example, racism may be useful to business owners as a way of justifying low pay for minority workers; in addition, racist attitudes and practices can also promote internal dissension and conflict within the proletariat, since workers may split along racial or ethnic lines in their competition for scarce jobs. Such divisions tend to exist inside nations, but they can also arise between nations in the larger world economy. For example, the predominantly white labour forces of powerful "core" countries may show little solidarity or identification with the problems of nonwhite workers in the less developed "peripheral" regions of the world (e.g., Bonacich, 1972, 1985; Wallerstein, 1989; Balibar and Wallerstein, 1991; Allahar, 1995; recall also Wright, 1994: 30).

Non-Marxist theorists may be less likely to attribute ethnic discord to the actions of the bourgeoisie, but these writers also adopt a *conflict* perspective to explain racial and ethnic inequality in many cases. The approach used in such analyses tends to be Weberian, with a focus on *collective action,* interest group *competition,* and differential access to *resources* as key mechanisms in the generation of ethnic inequality. Some of this research suggests that ethnic conflict and inequality occur not only when times are hard and jobs are scarce, as many Marxists argue, but also when prospects are improving. In the latter situation, ethnic groups may clash over who is to benefit from the new opportunities and prosperity (Olzak and Nagel, 1986; Belanger and Pinard, 1991; Olzak, 1992; recall also Parkin, 1979: 63–68).

Another sign of the Weberian influence on studies of race and ethnicity is the number of prominent contributors to this area who employ an approach built primarily on the role of power or domination. In Canada, the best illustration is probably John Porter's classic study, *The Vertical Mosaic.* This analysis conceives of ethnic (and class) inequality mainly in terms of the relative control that different groups are able to acquire within the country's economic, political, and ideological power structures (Porter, 1965). As we found in the discussions of gender inequality, here too there can be disagreements on the relative role of each of these three sources or types of power. For example, some Marxist writers are more likely than Porter to argue for the importance of economic power over all others. A related contention is that ethnic and racial inequalities are best conceived of as secondary fractions within a general system of social inequality that is "primarily based on class relations" (Li, 1988: 132, 140; but see Li, 1992: 489). Even these debates, however, arise within a context of general agreement on the importance of all three sources of power in creating and sustaining racial and ethnic inequalities (e.g., Ramcharan, 1982: 2–4, 97–98; Bolaria and Li, 1985: 1–11; Li, 1988: 23; 1992; for some discussion, see Satzewich, 1998; Bolaria, 2000; Henry et al., 2000; Fleras and Elliott, 2003).

Many of the best-known analyses of racial and ethnic inequality in the United States reveal a similar reliance on the concept of power and its various forms (e.g., Shibutani and Kwan, 1965; Blalock, 1967, 1989, 1991; Noel, 1968; Blauner, 1972; Wilson, 1973; for some discussion, see Pincus and Ehrlich, 1994; Hurst, 1995; Bonilla-Silva, 1997). Especially in his earlier writings, William Wilson conceived of black subordination in American society mainly as a problem stemming from limited access to various "power resources," including economic, political (or coercive), and ideological (or cultural) components (Wilson, 1973: 5, 7, 16–18). Although his emphasis on power has been supplanted to some extent by a more recent focus on class, the conceptual relevance of power for explaining racial inequality is still apparent (see Wilson, 1978, 1987, 1991, 1999).

In fact, our review of all the contemporary perspectives suggests an increasing agreement in much of this literature that, at a general level of analysis, the problem of social inequality is essentially a question of which groups have greater access to or control over the major sources of power. A related trend in recent analyses is the growing belief that each potential basis for inequality in society can be fully understood only if it is examined as part of a larger pattern of relations that involves all the others. Thus, as noted earlier in the discussion of gender perspectives, many leading writers now recognize that gender, race, and class inequalities often intermingle in significant and complex ways. This realization has led one theorist to assert that, in conceptual terms, "class, race, and gender are the *same* things—that is, their inferiorization is produced by the same theoretical logic" (Agger, 1993: 2; see also Connell, 1987: 292–293; Andersen and Collins, 1995: xi; Walby, 1997: 98).

The position we have taken throughout the present analysis is quite similar to this view. In other words, to the extent that race, gender, and class inequalities can each be understood to arise out of core power differences and related structures of domination, they are all, in that sense at least, the same thing. In truth, inequalities based on a wide range of social criteria can be perceived in this way, as intersecting elements in an overall system of domination and power inequity in society. Much the same perspective has prompted Margaret Andersen and Patricia Collins (1995: 5) to suggest that, in addition to race, class, and gender, such criteria as "age, religion, sexual orientation, physical ability, region, and ethnicity" can all be included in a general matrix of domination and inequality in modern societies.

GLOBALIZATION AND THEORIES OF SOCIAL INEQUALITY

A topic that has been of increasing interest to social scientists in recent years is the global or international nature of social inequality. Much of the discussion has been connected to the concept of "globalization," which, as Gordon Laxer has noted, (2004: 31) is now one of the most widely used terms in political, academic, and public discourse. However, as Laxer also makes apparent, concerns over globalization and the study of global inequality have a longer history than some current observers seem to assume (Laxer, 2004: 32).

Indeed, at least since the time of Marx, leading social theorists have sought to understand social inequality as an international or global problem. As we discussed in Chapter 2, for example, Marx and Engels believed that the historical spread of capitalist economic relations, which linked the industrialized and imperialist nations of the world to the less developed and colonized countries, was basic to capitalism's continued expansion and eventual transformation into socialism. In Marx's view, these worldwide processes of economic and social change would ulti-mately make clear the universal antagonism and polarization of class interests fundamentally dividing all workers from all owners on the planet. It was not Marx's purpose to present a detailed and systematic analysis of the international forces that were at work in this regard. Nevertheless, it is clear that he believed the proletarians of the world would in time come to see that national differences are unimportant, and that workers "have no country" (Marx and Engels 1848: 90). In *The Communist Manifesto*, Marx also leaves no doubt about the cause and the solution to global inequality, arguing that "workers of all countries" should unite, for they "have nothing to lose but their chains" and "have a world to win" (Marx and Engels, 1848: 116).

We have determined previously that, among the classical writers, the principal alternative to the Marxian class-based perspective on social inequality is Weber's power-based approach. It is of note that Weber also shows a concern with global patterns of inequality, although, like Marx, he does not offer a detailed assessment of this question. As we would expect from his more general perspective, Weber's views are somewhat different from Marx's, emphasizing that global inequality stems from a more pluralist set of power structures that go beyond economic power alone and that involve a more general system of rationalized and bureaucratized social hierarchies. Weber's views on global inequality are certainly consistent with those of Marx in noting the importance of economic power differences between classes (and other interest groups) that are engaged in competition for the control of external international markets (e.g., Weber, 1923: 208–213; see Collins, 1980: 927–929, 936–938). In addition, however, Weber's discussion of inequality among the world's nations is more "geopolitical" in nature (e.g., Collins, 1988b: 131–135). Weber stresses that, apart from competing economic forces, we must recognize the crucial impact of political and military factors—especially the use of coercive force by different countries and the role of modern bureaucratized nation-states—if we are to understand fully how some societies have succeeded in attaining advantages over, or domination of, other societies (e.g., Weber, 1923: 249–260).

Subsequent to Marx and Weber, alternative approaches to the study of global inequality have emerged. Following a pattern roughly parallel to the general trend depicted in Figure 6.1 at the beginning of this chapter (page 117), some writers associated with the structural-functionalist school, including Talcott Parsons, brought into prominence a quite different set of perspectives for explaining or understanding inequality among the world's nations. These perspectives fall under the general rubrics of "modernization" theory or "evolutionary" theory. Rather than emphasize the role of economic, political, or military power in accounting for global inequality, these approaches typically have pointed to ideological or cultural factors, especially national differences in commitment to "modern" values such as individual "achievement motivation" and the desire for economic progress and development (e.g., Parsons, 1966: 113–114; Lipset, 1967: 33; Inkeles and Smith, 1974; for discussion, see Baer et al., 1996: 300–304; Adams and Sydie, 2001: 357–358; Lenski, 2005: 131–133). According to these perspectives, some of the main reasons why the more prosperous and powerful societies, especially the United States, now enjoy the advantages that they do in the world can be traced to their stronger belief in and promotion of such values. The implication is that, if other countries would follow the lead of the more modern or more "socially evolved" societies, they too could attain comparable levels of affluence and influence within the family of nations.

With a few exceptions, more recent theorists have been highly critical of theories based on assumptions of modernization or societal evolution. As with the rejection of structural functionalism that, as we have discussed, occurred during the period after World War II, various theorists of international relations reacted against such claims about the reasons behind and the virtues of globalized capitalist development. Probably the most notable of these early critics was Andre Gunder Frank (1966, 1967), who argued that the economic inroads into the less developed nations of the world by the more affluent and powerful industrialized societies simply made matters worse for the poorer countries (for discussion, see, e.g., Allahar, 1995). The view of Frank and others was that this process mainly ensured that the less developed nations would remain underdeveloped and increasingly subordinate to the powerful capitalist interests operating out of the developed societies.

Since Frank's early analyses, the most influential work on the topic of global inequality has been Immanuel Wallerstein's "world-systems" approach (e.g., 1974, 1979, 1989, 1995; also Balibar and Wallerstein, 1991). Building from what is a Marxist perspective in some respects, Wallerstein has argued that, over the past five centuries, the world has moved away from a more multifaceted and diverse international system of competing capitalist economies, and toward what is essentially just one overarching global capitalist system (e.g. Wallerstein, 1974: 4; 1979: 133–134; see also Adams and Sydie, 2001: 457–459). At the same time, though, as Randall Collins has argued, it seems clear that "Wallerstein's revision of Marxism is in many ways a movement toward a more Weberian mode of analysis, stressing the importance of external relations among states" and not just capitalist economic influences (Collins, 1980: 938). Giddens has likewise noted that Wallerstein's approach incorporates Weberian ideas, including some recognition of the military and political roles played by nation-states in shaping the world economic system; even so, Giddens has criticized Wallerstein's analysis for understating the considerable powers that these military and political resources give to national governments, powers that can operate apart from the economic interconnections linking the capitalist systems or capitalist classes of different countries (e.g., Giddens, 1985: 166–171).

For the most part, however, it is evident that the leading theorists of global inequality, including Frank and Wallerstein, are broadly similar to the leading general theorists of inequality during the contemporary period, in that their key contributions also build upon fundamental insights to be found in the works of both Marx and Weber. This assertion, of course, is equally true of Giddens himself, who is the one recent general theorist that we have assessed who has written most extensively on the problem of global inequality. As we discovered previously in this chapter, Giddens draws upon key ideas about the nature of capitalism and state power that are evident in Marx's and Weber's writings, and extends these ideas in developing his own theoretical arguments about the problem of international inequality, especially the "global bads" that the world's peoples must now confront as we move through the twenty-first century.

CONCLUSION

With this brief review of the linkages between the leading general perspectives on social inequality and the more specific analysis of global inequality, we come to the end of our assessment of major contemporary theories. In the closing chapter, the central contributions of all the leading theorists will be summarized and briefly reconsidered. We will then offer an outline of the overall direction that past and present theories seem to be taking us in understanding the general problem of social inequality.

CHAPTER SEVEN

Theories of Social Inequality: An Overview and Evaluation

In this concluding chapter, it is important to restate and comment on the most crucial contributions and common threads that we have found in examining the various classical and contemporary perspectives on social inequality. Our discussion will be somewhat selective, for it is intended primarily to underscore the central insights and issues that have been raised, rather than to repeat in detail all of the themes and arguments that have been considered in previous chapters. This assessment will enable us to offer some concluding observations on what key elements should be included in a theoretical overview of social inequality in contemporary societies. We will also illustrate the feasibility of applying this general overview for the conceptualization of inequality in its more specific forms and manifestations.

THE MAJOR PERSPECTIVES: A SELECTIVE REVIEW

It should be clear at this stage that Marx's ideas, with which we began our analysis, still stand at the centre of any complete discussion of structured social inequality. Despite the many criticisms levelled against Marx by friend and foe alike, it seems certain that his understanding of the origins of class relations in the sphere of production, and his grasp of the internal workings and contradictions of capitalism, will continue to be lasting contributions to modern social thought. Marx's work is incomplete as a general perspective on social inequality, in part because the development of such a perspective was not his purpose. Nevertheless, in examining the mechanisms for the revolutionary overthrow of capitalism, the task that required most of his intellectual efforts, Marx also succeeded in identifying what many still believe is the principal cleavage, the great divide, in all systems of social inequality: between those who own or control the means of material production and those who survive through their labour power. Even in these times of socialist disarray, Marx's essential arguments about class and capitalism still provide the point of reference and departure for most of the writers who have followed him.

The second great figure in the study of social inequality is Max Weber. Although several important differences exist between Weber and Marx, there are also fundamental similarities in their perspectives, especially on the question of class and its role in the structure of modern society. Those who stress the disagreements in the ideas of Marx and Weber often seem to overlook the affinities between them, as well as Weber's self-professed wish to provide a positive and constructive, rather than a negative and destructive, critique of Marxist thought. Many of Weber's major contributions centre on this constructive effort to amplify or modify Marx. These contributions include: his recognition in the class system of positions that are neither bourgeois nor proletarian but tend to persist in a middle range between the two main classes; his perception that class relations are best understood as the key type among a more general or plural set of power relations, involving competing interests and multiple struggles among groups; his delineation of the numerous forms of domination in social structures, including not only legitimate power or authority but also the domination that derives from tradition, habit or custom, fear of repression, and so on; his identification of bureaucratic organizations as the predominant mechanisms for domination in the political, economic, and other spheres of modern life; and his early awareness of the increasing role played by the state in the overall power structure, through its monopoly of repressive force and its connections to the economic system. All these points add immeasurably to the subsequent assessments of social inequality in recent times.

The third classical theorist we considered was Durkheim. The inclusion of Durkheim in an analysis of inequality may be surprising to some, since he is not conventionally seen as a key figure in the field. However, Durkheim's thought offers us an important link between certain basic ideas in Marx and Weber and more recent structural-functionalist formulations. In particular, Durkheim reveals an awareness of such problems as class struggle and alienation under capitalism that is similar to Marx's view in some respects. At the same time, such Weberian concerns as the state and the role of power in society receive more than passing attention. Like Weber, Durkheim sees power, and the social constraints that go with it, arising out of a growing set of legal rules that nonetheless are supplemented at all times by the influence of informal traditions, conventions, habits or customs, and occasionally by coercion. Durkheim also anticipates more recent thinkers such as Giddens, in a sense, for he too notes the two-sided nature of power, that power can be a positive as well as a limiting force in society, and that rules not only establish power but also prevent its abuse. What separates Durkheim from Marx and Weber, and provides his otherwise tenuous connection to structural functionalism, is his greater stress on the positive, not the negative, consequences of power differences and inequality, specifically within the division of labour.

After Durkheim, we then examined the structural-functionalist school. Among structural functionalists, the predominant strategy is to emphasize Durkheim's sense of the integrative benefits of inequality and to ignore, for the most part, his awareness of the exploitation and injustice that also tend to exist in social hierarchies. This narrow reading of Durkheim is compounded by a peculiar interpretation of Weber, in which any inequalities and power relations that are not based on legitimate authority or consensus go largely unrecognized. The deficiencies in the structural-functionalist approach, which occur because of these omissions, have been discussed earlier. However, in spite of these and other weaknesses, there are at least two elements in structural-functionalist analyses that seem to have a sustained influence in the study of social inequality. First, as Giddens notes, structural functionalism has contributed greatly to the identification of major social institutions and the concrete economic, political, and other structures to which they are tied. Although structural functionalists rarely view them in this way, these structures are a useful step toward the delineation of a general set of structures of domination in society.

Second, the contention by structural functionalists that there is something inevitable and natural about social inequality is a view that a majority of the population advocates, especially when inequality refers to things like differential rewards for differences in talent or effort (e.g., Jasso and Rossi, 1977; Robinson and Bell, 1978; Della Fave, 1980; Kelley and Evans, 1993; Kluegel and Bobo, 1993). Of course, popular acceptance of inequality is not a confirmation that it is a natural or inevitable phenomenon. Moreover, as noted in Chapter 1, that social inequality exists in virtually all known societies is not a proof that it must persist in the future. Still, it is difficult to agree with those who simply explain away inequality as some doomed remnant of bourgeois consciousness, a false or distorted form of human relationship that will end with the demise of capitalism. The ready evidence in past and present socialist systems of unequal rights, opportunities, rewards, and privileges is sufficient in itself to cast doubt on such claims (e.g., Yanowitch, 1977; Lane, 1982; Giddens and Held, 1982; Giddens, 1994, 2000a). Hence, until we can achieve some universal shift to an altruistic consciousness among people, there would seem to be at least a kernel of truth in the structural-functionalist view of this question. Even subsequent critics of structural functionalism often appear to agree that such traits as effort, motivation, talent, or training are at least partly responsible for inequality and that, within certain limits, most people will perceive this situation as both inescapable and, to some extent, justifiable.

The acknowledgment of these insights and others may partly account for the resurgence of interest in structural functionalism within some academic circles. This new interest seems most apparent in the case of Parsons, whose work has received renewed praise and recognition by such prominent writers as Alexander, Giddens, and Habermas (Antonio, 1989; Johnson, 1993; see also Alexander, 1998: 12–13, 55).

Nevertheless, in spite of these considerations, there is little doubt that the structural-functionalist approach, particularly to the study of social inequality, has serious deficiencies. A key difficulty is that *individual* factors like talent and motivation, which are emphasized in the structural-functionalist explanation of inequality, seem often to be relatively minor influences, at least in comparison with the *structural* differences in power or class that exist in society. In addition, as some Marxists contend, whatever mobility occurs within these structures because of personal "worth" or effort does little or nothing to change the structures themselves. The point to stress is that established structures tend to define the prospects and life chances of people in most instances, although the roles of individual capacity and human agency should not be overlooked.

After structural functionalism, the first of the more recent theorists we considered was Dahrendorf. Dahrendorf's contributions to the study of inequality mainly involve his rejection of some of the more extreme elements in early structural-functionalist analysis. He is most notable for questioning the assumption that all forms of inequality are somehow based on consensus, for calling attention to the inherent conflict in social hierarchies, and for attempting to draw a conceptual connection between class and power (or more precisely, authority). Unfortunately, as we have seen, some of Dahrendorf's original work confuses matters by treating all authority differences as class differences. Even so, Dahrendorf continues to be one of the most eloquent and persuasive advocates of the view that social inequality in itself is not inherently evil, provided it is based on real opportunities for all people to develop their talents and to pursue their interests as citizens in a free society.

The next writer to be assessed, Lenski, is similar to Dahrendorf in some respects. First, Lenski's approach also fosters some confusion of the ideas of class and power. In addition, as in Dahrendorf's early work, Lenski's original analysis tends to overestimate the extent to which inequalities in class and power have been reduced in the modern era. However, Lenski more than compensates for these problems with his important identification of the multiple bases for power that arise in contemporary societies, and with his suggestion that power in its various forms is primarily responsible for the unequal distribution of material privilege and prestige to groups and individuals. In this manner, Lenski directs us to a more global conception of what social inequality is. In doing so, he also moves us back toward Weber, Marx, and the continuing debate between neo-Weberian and neo-Marxist views on the problem of social inequality.

A central conclusion of this book is that most of the promising developments in the general conceptualization of social inequality are still to be found in the disputes and discussions growing out of the Marxist class analyses of writers such as Wright and Poulantzas, on the one hand, and the power- or domination-based perspectives of theorists such as Giddens and Parkin, on the other hand. It should also be evident from the previous analysis that the overall thrust of our review generally favours the broadly Weberian views of the latter two writers. The reasons for this choice centre mainly on the greater generalizability of the power concept over the class concept and on the multiple forms of inequality that this allows us to recognize. The existence of some degree of pluralism, and of distinct bases for power in addition to class power, is *explicitly* denied by Poulantzas and some other Marxists; yet, as we have seen, these elements are *implicitly* acknowledged in the Marxist delineation of fractions or contradictory locations in the class system and of the economic, political, and ideological apparatuses in capitalist society. The reluctance to give express recognition to this pluralism, or to see a conceptual equivalence in the fact that class, gender, race, and other bases for inequality all involve power relations, may stem from a fear among some Marxists that class would thereby be diminished in importance as a concept in this area. Such a fear, however, is unwarranted. It is possible to arrive at a conception of social inequality that maintains a pivotal role for class analysis but that also appreciates the complex nature of both class relations and the other power relations that are at work in shaping social structures (see Miliband, 1987: 328; Edgell, 1993: 122).

In his most recent analyses, Wright stands out as one prominent Marxist who now agrees with this assessment. Not only does he reveal a greater willingness than many Marxists to recognize the pluralism of modern power structures, but he also acknowledges that bases for domination such as race and gender are in some cases more important than class in the overall pattern of inequality in society. Equally significant is Wright's early work, which continues to be invaluable for clarifying and redirecting issues first raised by Poulantzas and other Marxists. In particular, Wright's conception of the capitalist class structure, especially in its original form rather than its more recent guises, is arguably the most systematic and effective scheme available for portraying class relations in contemporary societies.

For Wright's approach to provide an even more complete picture, however, it also requires the incorporation of Giddens's duality of social arrangements. In this way, the skeleton of class positions delineated by Wright can be more thoroughly invested with the human content that both Giddens and Parkin favour when discussing classes. In that event, we can accept that relations of productive control form the underpinnings of the class system, but we can also affirm that clusters of real people will coalesce around the different bases for interaction or closure that are inherent to these relations (for some debate, see Burris, 1987; Giddens, 1989; Wright, 1997).

Such a dual sense of classes as structures and classes as people is useful for avoiding fruitless debates that insist classes must be only one or the other. Similarly empty controversies, over whether power is a capacity of persons or a relation between them and whether the state has power or not, could also be eased if it were seen that both actors and the patterns of interaction among social positions are involved in all these things. Although various writers reveal an awareness of this dual nature of class and power, Giddens has probably been most successful in incorporating it into his overall perspective on inequality in social systems. His treatment of power is particularly valuable for its assimilation of elements from several other analyses. Included here are his revitalized version of Weber's conception of domination; his use of institutional or structural elaborations similar to those in functionalism and recent Marxism; and his awareness, like Parkin, of the multiple forms of closure and exploitation in society, which contribute not only to class inequality but also to inequalities based on gender, race, and a whole range of other factors.

In our review of recent theories dealing with gender and race/ethnicity, and also in our brief treatment of perspectives focusing on the problem of global inequality, we found that many of the best writers specializing in these areas place a similar emphasis on the importance of different forms of power and domination for understanding social inequality. The growing literature on gender and race, in particular, is notable for extending ideas and filling in certain gaps left by the classical theorists and other writers. Another important contribution in recent work on gender and race is that it has added significantly to our awareness of the interconnectedness among all the major bases for inequality that exist in society.

SOCIAL INEQUALITY: A SUMMARY PORTRAIT

Having highlighted the crucial themes in classical and contemporary theories of inequality, we come again to the fundamental questions raised in the opening chapter. We are now in a better position to answer questions about what social inequality is and how it should be conceived; however, it is also clear that no single set of answers will be universally acceptable, since no single perspective can subsume the others or resolve all the disagreements among theorists. For these reasons, no grand synthesis of viewpoints will be attempted here.

Nevertheless, as we have seen, there are several fundamental insights that many, if not most, of the writers do share, especially at a general level of discussion. There should be no surprise in this; given the erudition of the thinkers involved, the surprising outcome would be if they did not agree on certain key points. It can be argued that the broad similarities found in many of the perspectives may also move us toward a "common vocabulary," as Parkin has called it, for discussing social inequality (Parkin, 1979: 42).

As a step in this direction, Figure 7.1 (page 212) offers a summary picture or composite scheme of elements that should be included in the general conceptualization of social inequality in contemporary societies. This depiction is essentially an abstract representation of the major means for establishing power relations in social settings, the resulting structures of domination that emerge and are reproduced in social interaction, and the principal bases for social inequality that typically operate within and across these structures.

Power and Domination

First, it is apparent that the concept of power is central to this portrait. In fact, while class and power are the two key concepts in most of the general perspectives on inequality we have reviewed, Figure 7.1 should make it clear that in this summary portrait, power is employed as the most common and pervasive conceptual element. Power is defined here as a differential capacity to command resources, which gives rise to structured, asymmetric relations of domination and subordination among social actors. There are three key means by which power is normally generated in social systems: control of material resources, of people, and of ideas. The first two means of power correspond roughly to Giddens's concepts of allocation and authorization and to a broadly similar delineation of "human and non-human resources" found in Goldthorpe (1974: 218; see also Grabb, 1982). The third designation of power through ideas is an extension of the twofold scheme. It is reminiscent of Francis Bacon's famous remark that "knowledge is power," that ideas and information can serve as the means, or the medium, of power in the same way that money or people can. The recognition of this third type of power is generally compatible with the work of previous writers we have examined, virtually all of whom, in varying degrees, note the importance of controlling special knowledge and information or the influence that beliefs and ideas can have on social life. In addition, by combining this category of ideological power with the other two types, we arrive at a classification that corresponds with the three-way division of social structures (economic, political, and ideological) that most of the theorists we have considered seem to advocate.

Figure 7.1
The Major Means of Power, Structures of Domination, and Bases for Social Inequality

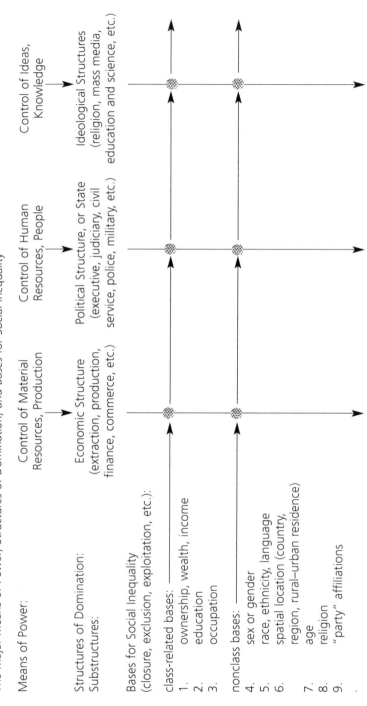

Thus, below the three means of power shown in Figure 7.1 are three attendant structures: the economic, the political, and the ideological. These social structures could be analyzed in many ways, but because of our specific interest in power and patterned relations of inequality, they are presented here as structures of *domination*. The connections in the diagram are oversimplified in that they imply a simple one-to-one linkage between each type of power and each structure. In fact, the downward arrows indicate only the *primary* linkages, since each form of power can operate in at least a secondary fashion in any of the structures. For example, control of people may be the principal activity within the political structure, but it is also evidenced in the control imposed by owners on workers in the economic structure. Similarly, the political structures that form the state do not derive all their power from the capacity to legislate or coerce human behaviour, for they also control material resources through tax revenues, government ownership of some business enterprises, and so on. As for control of ideas or knowledge, this is most clearly identified with the ideological structures, but it is also a means for retaining power elsewhere, as illustrated by the access to special technical knowledge in the economic structure and the information-gathering powers within the state.

In this manner, then, the vertical dimension in Figure 7.1 signifies the structures (and internal substructures) of domination in the social systems that form society. In concrete terms, these structures are manifested in various organizations or collectivities that operate at different levels of analysis. Many of the most crucial illustrations are large-scale or macroscopic in nature, including government bureaucracies or giant business corporations. In addition, as writers such as Chafetz and Walby have suggested, the organizations involved in systems of power and domination can be intermediate in size or social location, as in community groups or voluntary associations, and they can also be small-scale or microscopic entities, such as families, friendship groups, or even couples. Virtually all of these organizations and collectivities are patterned according to formal rules, laws, or rights of office, and most or all of them are also guided by a mix of informal practices, traditions, customs, or habits. These formal and informal rules and practices operate in what are often intricate and complex ways. Together, they largely determine the nature and extent of inequality in society, what the social bases for inequality will be and how much they will matter.

The Social Bases for Inequality

The horizontal dimension in Figure 7.1 provides a list of individual and group characteristics that typically have been used as the major bases for inequality in advanced societies. We have noted several times before the dualistic view of power fostered by various writers, most particularly Giddens. One way to look at Figure 7.1 is to think of the vertical dimension as the *structural* part of this duality and to treat the horizontal plane as the second part of the duality, the *human content* of power relations. Along this side of the diagram are the personal or socially defined attributes and affiliations that invest the structures of domination with their social and human elements. These attributes or affiliations are essentially similar to what one theorist has labelled "systacts" (Runciman, 1989: 20). They serve as the crucial factors around which group interests coalesce and collective action is organized, in the various competitions or struggles that can develop over access to the three different means of power. In combination, both the structural and the human components of power relations give rise to the multifaceted system of inequality that society comprises, with people processed and located at different levels, depending on the relative share of major economic, political, and ideological resources at their disposal.

Class-Related Bases for Inequality

The personal or socially defined attributes, capacities, or criteria for affiliation that are listed in Figure 7.1 can be divided into two main categories. First are three class-related factors: ownership (which also subsumes possession of wealth or income), education, and occupation. Ownership corresponds loosely with the various forms of property ownership that both Marxists and Weberians consider the key mode of economic control or closure in society. Education parallels the credentials basis for inequality that Parkin, Giddens, Wright, and others have stressed. Occupation subsumes such traits as skill level, manual versus nonmanual labour, and so on, that are not captured in the other two class-related bases for inequality. As some recent theorists have argued, occupation remains a pivotal element in any complete conceptualization of contemporary class structures (e.g., Grusky and Sorensen, 1998). These three factors operate within and across all the structures of domination shown in the vertical dimension. In conjunction, they push capitalist societies toward a class system that is highly complex, that varies in specific details from country to country and even from region to region, but that has at its core three major categories: an upper class composed mainly of the bourgeoisie, large-scale owners and controllers of material production, along with a smaller number of leaders from other spheres, who control political and ideological production and also tend to have sizable personal fortunes or business holdings; a heterogeneous central category of people who may have limited powers of ownership, such as the ownership of small businesses or financial investments, but who are distinguished in most cases by their access to special education, formal training, recognized skills, or other credentials; and a lower or working class whose members are largely without property or special credentials and rely almost completely on labour power for their material existence (see also Grabb, 2004: 6–7).

Although this image of the class structure seems closest to that of Weber or Giddens, it is also similar to that posed by Marxists such as Wright. Until recently, the most contentious issue has been whether the central category represents a separate middle class all on its own or, instead, a set of contradictory or dual locations that remain secondary to the two main classes. It is interesting that, at least in Wright's case, he now readily refers to these intermediate positions as "the middle class," in spite of the non-Marxist or Weberian connotations of that term. It is worth noting, as an aside, that Marx himself also had no difficulty using the "middle class" label for such intermediate positions at times (Grabb, 2004: 5). In the end, though, the choice of terms may be less important than the simple truth that there is something else in addition to the capital-versus-wage-labour distinction that exists in reality and requires explanation or study. That both Marxists and non-Marxists now tend to agree on this third category, if not always on how to label it, seems sufficient grounds for including it here.

Nonclass Bases for Inequality

The second set of human attributes or affiliations shown in the horizontal dimension of Figure 7.1 includes bases for inequality that are not inherent to class formation, although they are normally correlated with class inequalities in most cases. Whether and to what degree these nonclass bases for inequality actually operate will clearly differ from society to society. As we have seen, leading non-Marxists, such as Giddens and Parkin, along with prominent Marxists like Wright, agree that the relative significance of class and nonclass factors in the overall structure of inequality is likely to vary across different situations and times. Many of the most influential perspectives on gender and racial inequality take a similar view, and, as we have observed, they also argue that a thorough analysis of both class and nonclass inequalities should consider their probable interconnections and interactions.

It should be noted that the list of nonclass bases for inequality shown in Figure 7.1 may not be exhaustive. There are other traits or affiliations that are less obvious or less commonly recognized in the literature as important determinants of people's life chances but that, nevertheless, can play a role in the structure of social inequality. These include physical beauty or disability, for example. Generally, though, the bases listed here appear to be the major factors operating within modern systems of inequality.

Most of the bases for inequality listed in Figure 7.1 have been mentioned in one way or another by the majority of classical and contemporary theorists we have reviewed. However, Giddens is one of the few writers we have focused on who makes direct reference to the ways in which geographic location can affect patterned relations of inequality. Giddens's key example is the exploitation of one nation-state or country by another. As we have seen, this form of inequality has been tied to the increasing globalization of capitalism in recent times, and to the processes by which developed nations systematically keep less-developed nations in situations of subordination in the world economy (recall Frank, 1966, 1967; Wallerstein, 1974, 1979, 1989; Balibar and Wallerstein, 1991; Allahar, 1995; for recent research, see Breton and Reitz, 2003; Carroll, 2004; Milanovic, 2005). In addition, however, the geographic inequalities that arise *within* countries can also be delineated, especially between people of different regions or between those who live in rural versus urban settings. These spatial bases for inequality are not recognized as significant by all sociologists, although Canadian researchers have been interested for some time in the problems of regional disparities in economic development and in the more general inequalities involved in "core–periphery" or "metropolis–hinterland" relations (e.g., Innis, 1956; Creighton, 1956; Davis, 1971; Matthews, 1983; Bryan, 1986; Coffey and Polese, 1987; Brodie, 1990).

The final basis for inequality listed in Figure 7.1 concerns access to or exclusion from the power that people experience through their party affiliations. Following Weber, this refers not just to formal political parties. It also serves as a residual term that subsumes all the other organized collectivities or interest groups that people may create in the general contest for power in social settings: trade unions, professional associations, public-interest organizations, voluntary associations, various pressure groups, and so on.

It is no coincidence that virtually all of the bases for inequality that are summarized in Figure 7.1 have served as rallying points for collective action among those seeking to reduce or eliminate social injustice in the contemporary period. In fact, it could be argued that, by their very nature, theories dealing with social inequality are almost always "emancipatory" (see, e.g., Balibar and Wallerstein, 1991; Wright, 1994). Thus, from Marx's time to the present day, most writers in this field have sought not only to explain how inequality arises but also to find the means to free people from its grip. Among the most prominent examples of these collective attempts to achieve greater equality are the American civil-rights movement, the drive for both francophone and Native people's rights in Canada, the fight against apartheid in South Africa, and the women's movement in a wide range of developed and less developed nations. While it has not been possible in this book to consider these and other initiatives for social change in detail, such activities clearly illustrate the continuing, and perhaps increasing, importance of the study of inequality for understanding the general problem of human rights in the modern world (e.g., Cairns and Williams, 1985; Carroll, 1992; Larana, Johnston, and Gusfield, 1994; Marx and McAdam, 1994).

CONCLUSION

With these remarks, our discussion and evaluation of the major perspectives on social inequality are now complete. In a sense, of course, the analysis of theory is never a finished project, but forever a work in progress. One of the main lessons of our review of leading writers is that any existing perspective, however significant, is continually subject to reconsideration or replacement with time. New ideas and refinements of established conceptions will regularly come to light and are essential for us to achieve a cumulative understanding of the social world. Ultimately, then, the most reasonable hope for the present analysis is that we have clarified the key insights of classical and contemporary theory as they currently stand. In doing so, we may also have painted, more boldly than before, the broad brushstrokes of a general portrait of social inequality.

References

Aberle, D.F., A.K. Cohen, A.K. Davis, M.J. Levy, Jr., and F.X. Sutton. (1950). The functional prerequisites of society. *Ethics* 60 (January), 100–111.

Acker, Joan. (1989). The problem with patriarchy. *Sociology* 23 (2) (May), 235–240.

Adams, Bert N., and R.A. Sydie. (2001). *Sociological theory.* Thousand Oaks, CA: Pine Forge Press.

Agger, Ben. (1992). *The discourse of domination.* Evanston, IL: Northwestern University Press.

Agger, Ben. (1993). *Gender, culture, and power.* Westport, CT: Praeger.

Alexander, Jeffrey C. (1983). *Theoretical logic in sociology: Vol. 3. The classical attempt at theoretical synthesis: Max Weber.* Berkeley: University of California Press.

Alexander, Jeffrey C. (Ed.). (1985). *Neofunctionalism.* Beverly Hills: Sage.

Alexander, Jeffrey C. (1987). *Twenty lectures: Sociological theory since World War II.* New York: Columbia University Press.

Alexander, Jeffrey C. (1990a). *Structure and meaning: Relinking classical sociology.* New York: Columbia University Press.

Alexander, Jeffrey C. (1990b). Differentiation theory: Problems and prospects. In J.C. Alexander and P. Colony (Eds.), *Differentiation theory and social change: Comparative and theoretical perspectives* (pp. 1–15). New York: Columbia University Press.

Alexander, Jeffrey C. (1998). *Neofunctionalism and after.* Oxford: Blackwell Publishers.

Alexander, Jeffrey C., and Paul Colomy (Eds.). (1990). *Differentiation theory and social change: Comparative and theoretical perspectives.* New York: Columbia University Press.

Allahar, Anton. (1995). *Sociology and the periphery* (2nd ed.). Toronto: Garamond.

Althusser, Louis. (1969). *For Marx.* New York: Pantheon.

Althusser, Louis. (1976). *Essays in self-criticism.* London: New Left Books.

Althusser, Louis, and Étienne Balibar. (1970). *Reading Capital.* London: New Left Books.

Andersen, Margaret, and Patricia Collins (Eds.). (1995). *Race, class, and gender.* Belmont, CA: Wadsworth.

Antonio, Robert J. (1989). The normative foundations of emancipatory theory: Evolutionary versus pragmatic perspectives. *American Journal of Sociology* 94 (4) (January), 721–748.

Arai, Bruce. (1995). Self-employment and the nature of the contemporary Canadian economy. Unpublished doctoral dissertation, Department of Sociology, University of British Columbia, Vancouver, Canada.

Armstrong, Pat, and Hugh Armstrong. (1984). *The double ghetto* (Rev. ed.). Toronto: McClelland and Stewart.

Armstrong, Pat, and Hugh Armstrong. (1990). *Theorizing women's work.* Toronto: Garamond.

Aron, Raymond. (1970). *Main currents in sociological thought* (Vol. 2). Garden City: Anchor Books.

Avineri, Shlomo. (1968). *The social and political thought of Karl Marx.* Cambridge: Cambridge University Press.

Baer, Douglas, James Curtis, Edward Grabb, and William Johnston. (1996). What values do people prefer in children? A comparative analysis of survey evidence from fifteen countries. In C. Seligman, J. Olson, and M. Zanna (Eds.), *The psychology of values: The Ontario Symposium, Volume 8* (pp. 299–328). Mahwah, NJ: Lawrence Erlbaum Associates.

Baer, Doug, Edward Grabb, and William Johnston. (1987). Class, crisis, and political ideology: Recent trends. *Canadian Review of Sociology and Anthropology* 24 (1) (February), 1–22.

Baer, Doug, Edward Grabb, and William Johnston. (1990). The values of Canadians and Americans: A critical analysis and reassessment. *Social Forces* 68 (3) (March), 693–713.

Balibar, Étienne, and Immanuel Wallerstein. (1991). *Race, nation, class: Ambiguous identities.* London: Verso.

Balkwell, J.W., F.L. Bates, and A.P. Garbin. (1982). Does the degree of consensus on occupational status evaluations differ by socio-economic stratum? Response to Guppy. *Social Forces* 60 (June), 1183–1189.

Baran, Paul, and Paul Sweezy. (1966). *Monopoly capital.* New York: Monthly Review Press.

Barber, Bernard. (1957). *Social stratification.* New York: Harcourt Brace and World.

Baxter, Janeen. (1994). Is husband's class enough? Class location and class identity in the United States, Sweden, Norway, and Australia. *American Sociological Review* 59 (April), 220–235.

Belanger, Sarah, and Maurice Pinard. (1991). Ethnic movements and the competition model: Some missing links. *American Sociological Review* 56 (August), 446–457.

Bell, Daniel. (1960). *The end of ideology.* New York: The Free Press.

Bell, Wendell, and Robert V. Robinson. (1980). Cognitive maps of class and racial inequalities in England and the United States. *American Journal of Sociology* 86 (September), 320–349.

Bem, Sandra. (1993). *The lenses of gender.* New Haven: Yale University Press.

Bendix, Reinhard. (1962). *Max Weber: An intellectual portrait.* Garden City: Anchor Books.

Bendix, Reinhard, and Guenther Roth. (1971). *Scholarship and partisanship: Essays on Max Weber.* Berkeley: University of California Press.

Benson, Leslie. (1978). *Proletarians and parties.* London: Methuen.

Berlin, Isaiah. (1963). *Karl Marx: His life and environment.* Oxford: Oxford University Press.

Bernard, Jessie. (1971). *Women and the public interest.* Chicago: Aldine.

Binns, David. (1977). *Beyond the sociology of conflict.* London: Macmillan.

Blalock, H.M., Jr. (1967). *Toward a theory of minority group relations.* New York: Wiley.

Blalock, H.M., Jr. (1989). *Power and conflict: Toward a general theory.* Newbury Park: Sage.

Blalock, H.M., Jr. (1991). *Understanding social inequality.* Newbury Park: Sage.

Blau, Peter. (1964). *Exchange and power in social life.* New York: Wiley.

Blau, Peter. (1977). *Inequality and heterogeneity.* New York: The Free Press.

Blau, Peter, and Otis Dudley Duncan. (1967). *The American occupational structure.* New York: Wiley.

Blauner, Robert. (1972). *Racial oppression in America.* New York: Harper and Row.

Blishen, Bernard. (1967). A socioeconomic index for occupations in Canada. *Canadian Review of Sociology and Anthropology* 4, 41–53.

Blishen, Bernard, William Carroll, and Catherine Moore. (1987). The 1981 socioeconomic index for occupations in Canada. *Canadian Review of Sociology and Anthropology* 24 (4) (November), 465–488.

Blishen, Bernard, and Hugh McRoberts. (1976). A revised socioeconomic index for occupations. *Canadian Review of Sociology and Anthropology* 13 (February), 71–79.

Blumberg, Rae Lesser. (1978). *Stratification: Socioeconomic and sexual inequality.* Dubuque: William C. Brown Company.

Blumberg, Rae Lesser. (1984). A general theory of gender stratification. In R. Collins (Ed.), *Sociological theory 1984* (pp. 23–101). San Francisco: Jossey-Bass.

Blumberg, Rae Lesser. (1989). Toward a feminist theory of development. In R. Wallace (Ed.), *Feminism and sociological theory* (pp. 161–199). Newbury Park: Sage.

Blumberg, Rae Lesser. (1991a). Women and the wealth and well-being of nations; macro-micro interrelationships. In J. Huber (Ed.), *Macro-micro linkages in sociology* (pp. 121–140). Newbury Park: Sage.

Blumberg, Rae Lesser (Ed.). (1991b). *Gender, family, and economy: The triple overlap.* Beverly Hills: Sage.

Blumberg, Rae Lesser. (2004). Extending Lenski's schema to hold up both halves of the sky: A theory-guided way to conceptualizing agrarian societies that illuminates a puzzle about social stratification. *Sociological Theory* 22 (2) (June), 278–291.

Blumberg, Rae Lesser, Cathy Rakowski, Irene Tinker, and Micheal Monteon (Ed.). (1995). *Engendering wealth and well-being.* Boulder, CO: Westview Press.

Bolaria, B. Singh (Ed.). (2000). *Social issues and contradictions in Canadian society.* Toronto: Harcourt Brace Canada.

Bolaria, B. Singh, and Peter S. Li. (1985). *Racial oppression in Canada.* Toronto: Garamond.

Bologh, Roslynn Wallach. (1987). Marx, Weber, and masculine theorizing. In Norbert Wiley (Ed.), *The Marx–Weber debate* (pp. 145–168). Newbury Park, CA: Sage.

Bonacich, Edna. (1972). A theory of ethnic antagonism: The split labor market. *American Sociological Review* 37, 547–559.

Bonacich, Edna. (1985). Class approaches to ethnicity and race. In N.R. Yetman (Ed.), *Majority and minority: The dynamics of race and ethnicity in American life* (pp. 62–77). Boston: Allyn and Bacon.

Bonilla-Silva, Eduardo. (1997). Rethinking racism: Toward a structural interpretation. *American Sociological Review* 62 (3) (June), 465–480.

Boswell, Terry, and William J. Dixon. (1993). Marx's theory of rebellion: A cross-national analysis. *American Sociological Review* 58 (October), 681–702.

Braverman, Harry. (1974). *Labor and monopoly capital.* New York: Monthly Review Press.

Breton, Raymond, and Jeffrey Reitz (Eds.). (2003). *Globalization and society: Processes of differentiation examined.* Westport, CT: Praeger.

Brodie, Janine. (1990). *The political economy of Canadian regionalism.* Toronto: Harcourt Brace Jovanovich Canada.

Bryan, Ingrid. (1986). *Economic policies in Canada* (2nd ed.). Toronto: Butterworths.

Burris, Val. (1987). The neo-Marxist synthesis of Marx and Weber on class. In Norbert Wiley (Ed.), *The Marx–Weber debate* (pp. 67–90). Newbury Park, CA: Sage.

Cairns, Allan, and Cynthia Williams. (1985). *Constitutionalism, citizenship, and society in Canada.* Toronto: University of Toronto Press.

Camic, Charles (Ed.). (1991). *The early essays of Talcott Parsons.* Chicago: University of Chicago Press.

Carchedi, Guglielmo. (1977). *On the economic identification of social classes.* London: Routledge.

Carchedi, Guglielmo. (1987). *Class analysis and social research.* Oxford and New York: Basil Blackwell.

Carroll, William. (2004). *Corporate power in a globalizing world.* Don Mills: Oxford University Press.

Carroll, William K. (Ed.). (1992). *Organizing dissent: Contemporary social movements in theory and practice.* Toronto: Garamond.

Cassell, Philip (Ed.). (1993). *The Giddens reader.* Stanford: Stanford University Press.

Chafetz, Janet Saltzman. (1984). *Sex and advantage: A comparative, macro-structural theory of sex stratification.* Totowa, NJ: Rowman and Allanheld.

Chafetz, Janet Saltzman. (1988). *Feminist sociology: An overview of contemporary theories.* Itasca, IL: F.E. Peacock.

Chafetz, Janet Saltzman. (1990). *Gender equity: An integrated theory of stability and change.* Newbury Park: Sage.

Chafetz, Janet Saltzman. (1997). Feminist theory and sociology: Underutilized contributions for mainstream theory. *Annual Review of Sociology* 23, 97–120.

Chafetz, Janet Saltzman. (1999). The varieties of gender theory in sociology. In J.S. Chafetz (Ed.), *Handbook of the sociology of gender* (pp. 3–24). New York: Klewer Plenum Publishers.

Clark, Terry Nichols, and Seymour Martin Lipset. (1991). Are social classes dying? *International Sociology* 6 (December), 397–410.

Clark, Terry Nichols, Seymour Martin Lipset, and Michael Rempel. (1993). The declining political significance of class. *International Sociology* 8 (September), 293–316.

Clement, Wallace, and John Myles. (1994). *Relations of ruling. Class and gender in postindustrial societies.* Montreal and Kingston: McGill-Queen's University Press.

Coffey, William, and Mario Polese (Eds.). (1987). *Still living together.* Halifax: Institute for Research on Public Policy.

Cohen, G.A. (1978). *Karl Marx's theory of history: A defense.* Princeton: Princeton University Press.

Coleman, Richard P., and Lee Rainwater. (1978). *Social standing in America: New dimensions of class.* New York: Basic Books.

Collins, Randall. (1975). *Conflict sociology: Toward an explanatory science.* New York: Academic Press.

Collins, Randall. (1979). *The credential society.* New York: Academic Press.

Collins, Randall. (1980). Weber's last theory of capitalism: A systematization. *American Sociological Review* 45 (6) (December): 925–942.

Collins, Randall. (1985). *Three sociological traditions.* New York and Oxford: Oxford University Press.

Collins, Randall. (1986a). *Weberian sociological theory.* Cambridge: Cambridge University Press.

Collins, Randall. (1986b). *Max Weber: A skeleton key.* Beverly Hills: Sage.

Collins, Randall. (1988a). The Durkheimian tradition in conflict sociology. In J.C. Alexander (Ed.), *Durkheimian sociology: Cultural studies* (pp. 107–128). Cambridge: Cambridge University Press.

Collins, Randall. (1988b). *Theoretical sociology.* San Diego: Harcourt Brace Jovanovich.

Collins, Randall. (1992). *Sociological insight: An introduction to non-obvious sociology* (2nd ed.). New York: Oxford University Press.

Collins, Randall. (1994). *Four sociological traditions.* New York and Oxford: Oxford University Press.

Connell, R.W. (1979). A critique of the Althusserian approach to class. *Theory and Society* 8 (3), 321–345.

Connell, R.W. (1987). Gender and power: Society, the person and sexual politics. Cambridge: Polity Press.

Coser, Lewis A. (1977). *Masters of sociological thought.* New York: Harcourt Brace Jovanovich.

Coxon, A., and C. Jones. (1978). *The images of occupational prestige.* London: Macmillan.

Creighton, Donald. (1956). *The commercial empire of the St. Lawrence.* Toronto: Macmillan.

Crompton, R., and J. Gubbay. (1977). *Economy and class structure.* London: Macmillan.

Crompton, Rosemary, and Michael Mann (Eds.). (1986). *Gender and stratification.* Cambridge: Polity Press.

Curtis, James E., Edward Grabb, and Neil Guppy (Eds.). (2004). *Social inequality in Canada: Patterns, problems, policies* (4th ed.). Toronto: Pearson Education Canada.

Dahrendorf, Ralf. (1958). Out of Utopia: Toward a reorientation of sociological analysis. *American Journal of Sociology* 64 (September), 115–127.

Dahrendorf, Ralf. (1959). *Class and class conflict in industrial society.* Stanford: Stanford University Press.

Dahrendorf, Ralf. (1968). *Essays in the theory of society.* London: Routledge and Kegan Paul.

Dahrendorf, Ralf. (1969). On the origin of inequality among men. In A. Beteille (Ed.), *Social inequality* (pp. 16–44). Middlesex: Penguin.

Dahrendorf, Ralf. (1979). *Life chances.* London: Weidenfeld and Nicolson.

Dahrendorf, Ralf. (1988). *The modern social conflict. An essay on the politics of liberty.* London: Weidenfeld and Nicolson.

Dahrendorf, Ralf. (1990a). No third way. *Partisan Review* 57 (4) (Fall), 508–525.

Dahrendorf, Ralf. (1990b). *Reflections on the revolution in Europe.* London: Chatto and Windus Ltd.

Dahrendorf, Ralf. (1995). A precarious balance: Economic opportunity, civil society, and political liberty. Cited in David Crane, Social cohesion victim of Harris's cuts. *The Toronto Star*, October 1, p. C2.

Dahrendorf, Ralf. (1997). *After 1989: Morals, revolution, and civil society.* Houndmills, Basingstoke: Macmillan.

Dahrendorf, Ralf. (2004). Why security depends on poverty and development. *Europaeum Review* 6 (1) (Spring), 20–23.

Davis, A.K. (1971). Canadian society and history as hinterland versus metropolis. In R.J. Ossenberg (Ed.), *Canadian society: Pluralism, change, and conflict* (pp. 6–32). Scarborough: Prentice-Hall Canada.

Davis, Kingsley. (1949). *Human society.* New York: Macmillan.

Davis, Kingsley. (1953). Reply to Tumin. *American Sociological Review* 18 (August), 394–397.

Davis, Kingsley. (1959). The myth of functional analysis as a special method in sociology and anthropology. *American Sociological Review* 24 (December), 757–772.

Davis, Kingsley, and Wilbert E. Moore. (1945). Some principles of stratification. *American Sociological Review* 10 (April), 242–249.

Della Fave, L. Richard. (1980). The meek shall not inherit the earth: Self-evaluation and the legitimacy of stratification. *American Sociological Review* 45 (December), 955–971.

Demerath, N.J., and Richard A. Peterson (Eds.). (1967). *System, change, and conflict.* New York: The Free Press.

Dunn, Dana, Elizabeth Almquist, and Janet Saltzman Chafetz. (1993). Macro-structural perspectives on gender inequality. In Paula England (Ed.), *Theory on gender/feminism on theory* (pp. 69–90). New York: Aldine de Gruyter.

Durkheim, Émile. (1893). *The division of labor in society* (1st ed.). New York: The Free Press, 1964.

Durkheim, Émile. (1895). *The rules of sociological method.* New York: The Free Press, 1964.

Durkheim, Émile. (1896). *Socialism and Saint-Simon.* Yellow Springs, OH: Antioch Press, 1958.

Durkheim, Émile. (1897). *Suicide.* New York: The Free Press, 1951.

Durkheim, Émile. (1902). *The division of labor in society. Preface to the second edition: Some notes on occupational groups.* New York: The Free Press, 1964.

Edgell, Stephen. (1993). *Class.* London: Routledge.

Eisenstein, Zillah. (1979). *Capitalist patriarchy and the case for socialist feminism.* New York: Monthly Review Press.

Elster, Jon. (1985). *Making sense of Marx.* Cambridge: Cambridge University Press.

Engels, Friedrich. (1872). On authority. In R.C. Tucker (Ed.), *The Marx–Engels reader* (2nd ed.) (pp. 730–733). New York: Norton, 1978.

Engels, Friedrich. (1882). Letter from Engels to Bernstein. In *Marx Engels Werke* (Vol. 35). Berlin: Dietz Verlag. Institut fur Marxismus-Leninismus Beim ZK Der Sed, 1967.

Engels, Friedrich. (1884). *The origin of the family, private property, and the state.* New York: International Publishers, 1972.

Engels, Friedrich. (1890a). Letter from Engels to Conrad Schmidt. In *Marx Engels Werke* (Vol. 37, pp. 488–495). Berlin: Dietz Verlag. Institut fur Marxismus-Leninismus Beim ZK Der Sed, 1967.

Engels, Friedrich. (1890b). Letter from Engels to J. Bloch. In *Marx Engels Werke* (Vol. 37, pp. 462–465). Berlin: Dietz Verlag. Institut fur Marxismus-Leninismus Beim ZK Der Sed, 1967.

England, Paula. (1992). *Comparable worth: Theories and evidence.* New York: Aldine de Gruyter.

England, Paula (Ed.). (1993). *Theory on gender/feminism on theory.* New York: Aldine de Gruyter.

Fallding, Harold. (1968). *The sociological task.* Englewood Cliffs: Prentice-Hall.

Fallding, Harold. (1972). Only one sociology. *British Journal of Sociology* 23 (March), 93–101.

Fenton, Steve. (1980). Race, class, and politics in the work of Émile Durkheim. In *Sociological theories: Race and colonialism* (pp. 143–182). Paris: UNESCO.

Fenton, Steve. (1984). Race and society: Primitive and modern. In S. Fenton with R. Reiner and I. Hamnett (Eds.), *Durkheim and modern sociology* (pp. 116–142). Cambridge: Cambridge University Press.

Firebaugh, Glenn, and Frank D. Beck. (1994). Does economic growth benefit the masses? Growth, dependence, and welfare in the Third World. *American Sociological Review* 59 (October), 631–653.

Fleras, Augie, and Jean Leonard Elliott. (2003). *Unequal relations* (4th ed.). Toronto: Pearson Education Canada.

Fox, Bonnie. (1988). Conceptualizing 'patriarchy.' *Canadian Review of Sociology and Anthropology* 25 (2) (May), 163–182.

Frank, Andre Gunder. (1966). The development of underdevelopment. *Monthly Review* 18, 17–31.

Frank, Andre Gunder. (1967). *Capitalism and underdevelopment in Latin America*. New York: Monthly Review Press.

Friedan, Betty. (1963). *The feminine mystique*. New York: Dell.

Gagliani, Giorgio. (1981). How many working classes? *American Journal of Sociology* 87 (September), 259–285.

Gerth, Hans, and C. Wright Mills (Eds.). (1967). *From Max Weber: Essays in sociology*. Oxford: Oxford University Press.

Giddens, Anthony. (1971). *Capitalism and modern social theory*. Cambridge: Cambridge University Press.

Giddens, Anthony. (1972). Politics and sociology in the thought of Max Weber. London: Macmillan.

Giddens, Anthony. (1973). *The class structure of the advanced societies*. London: Hutchinson.

Giddens, Anthony. (1976). *New rules of sociological method*. London: Hutchinson.

Giddens, Anthony. (1977). *Studies in social and political theory*. London: Hutchinson.

Giddens, Anthony. (1979). *Central problems in social theory*. Berkeley: University of California Press.

Giddens, Anthony. (1980). Classes, capitalism, and the state. *Theory and Society* 9 (November), 877–890.

Giddens, Anthony. (1981a). Postscript (1979). In *The class structure of the advanced societies* (2nd ed.) (pp. 295–320). London: Hutchinson.

Giddens, Anthony. (1981b). *A contemporary critique of historical materialism: Vol. 1. Power, property, and the state*. London: Macmillan.

Giddens, Anthony. (1982). *Profiles and critiques in social theory*. Berkeley and Los Angeles: University of California Press.

Giddens, Anthony. (1984). *The constitution of society*. Berkeley and Los Angeles: University of California Press.

Giddens, Anthony. (1985). *A contemporary critique of historical materialism: Vol. 2. The nation-state and violence*. Berkeley and Los Angeles: University of California Press.

Giddens, Anthony (Ed.). (1986). *Durkheim on politics and the state*. Cambridge: Polity Press.

Giddens, Anthony. (1989). A reply to my critics. In D. Held and J. Thompson (Eds.), *Social theory of modern societies* (pp. 249–301). Cambridge: Cambridge University Press.

Giddens, Anthony. (1990). *The consequences of modernity*. Stanford: Stanford University Press.

Giddens, Anthony. (1992). *The transformation of intimacy*. Cambridge: Polity Press.

Giddens, Anthony. (1994). *Beyond left and right. The future of radical politics.* Stanford: Stanford University Press.

Giddens, Anthony. (1998). *The third way. The renewal of social democracy.* Cambridge: Polity Press.

Giddens, Anthony. (2000a). *The third way and its critics.* Cambridge: Polity Press.

Giddens, Anthony. (2000b). *Runaway world. How globalization is reshaping our lives.* New York: Routledge.

Giddens, Anthony. (2002). *Where now for New Labour?* Cambridge: Policy Network.

Giddens, Anthony (Ed.). (2003). *The progressive manifesto. New ideas for the centre-left.* Cambridge: Policy Network.

Giddens, Anthony, and Patrick Diamond (Eds.). (2005). *The new egalitarianism.* Cambridge: Polity Press.

Giddens, Anthony, and David Held (Eds.). (1982). *Classes, power, and conflict.* Berkeley: University of California Press.

Goldstone, Jack A. (1986). State breakdown in the English Revolution: A new synthesis. *American Journal of Sociology* 92 (2) (September), 257–322.

Goldthorpe, John H. (1972). Class, status, and party in modern Britain. *European Journal of Sociology* 13, 342–372.

Goldthorpe, John H. (1974). Social inequality and social integration in modern Britain. In D. Wedderburn (Ed.), *Poverty, inequality, and class structure* (pp. 217–238). London: Cambridge University Press.

Goldthorpe, John H. (1983). Women and class analysis: In defence of the conventional view. *Sociology* 17, 465–488.

Goldthorpe, John H. (1984). Women and class analysis: A reply to the replies. *Sociology* 18, 491–499.

Goldthorpe, John H. (1987). *Social mobility and class structure in modern Britain.* Oxford: Clarendon Press.

Goldthorpe, John H., and Robert Erikson. (1992). *The constant flux.* Oxford: Clarendon Press.

Goldthorpe, John H., and Keith Hope. (1974). *The social grading of occupations: A new approach and scale.* Oxford: Clarendon Press.

Goldthorpe, John H., and Clive Payne. (1986). On the class mobility of women: Results from different approaches to the analysis of recent British data. *Sociology* 20, 531–555.

Grabb, Edward G. (1980). Marxist categories and theories of class: The case of working class authoritarianism. *Pacific Sociological Review* 33 (4) (October), 359–376.

Grabb, Edward G. (1982). Social stratification. In J.J. Teevan (Ed.), *Introduction to sociology: A Canadian focus* (pp. 121–157). Scarborough: Prentice-Hall Canada.

Grabb, Edward G. (2004). Conceptual issues in the study of social inequality. In J. Curtis, E. Grabb, and N. Guppy (Eds.), *Social inequality in Canada: Patterns, problems, policies* (4th ed.) (pp. 1–16). Toronto: Pearson Education Canada.

Grabb, Edward G., and Ronald D. Lambert. (1982). The subjective meanings of social class among Canadians. *Canadian Journal of Sociology* 7 (3), 297–307.

Grusky, David. (2005). Foundations of a neo-Durkheimian class analysis. In collaboration with Gabriela Galescu. In E.O. Wright (Ed.), *Approaches to class analysis* (pp. 51–81). Cambridge: Cambridge University Press.

Grusky, David, and Jesper B. Sorensen. (1998). Can class analysis be salvaged? *American Journal of Sociology* 103 (5) (March), 1187–1234.

Guppy, L. Neil. (1981). Occupational prestige and conscience collective: The consensus debate reassessed. Unpublished doctoral dissertation, Sociology Department, University of Waterloo.

Guppy, L. Neil. (1982). On intersubjectivity and collective conscience in occupational prestige research: A comment on Balkwell-Bates-Garbin and Kraus-Schild-Hodge. *Social Forces* 60 (June), 1178–1182.

Haas, Ain. (1993). Social inequality in aboriginal North America: A test of Lenski's theory. *Social Forces* 72 (December), 295–313.

Habermas, Jurgen. (1975). *Legitimation crisis.* Boston: Beacon Press.

Habermas, Jurgen. (1984). *The theory of communicative action.* Boston: Beacon Press.

Hagan, John, and Celesta Albonetti. (1982). Race, class, and the perception of criminal justice in America. *American Journal of Sociology* 88 (2) (September), 329–355.

Hagan, John, and Patricia Parker. (1985). White-collar crime and punishment: The class structure and legal sanctioning of securities violations. *American Sociological Review* 50 (3) (June), 302–316.

Halaby, Charles N. (1993). Reply to Wright. *American Sociological Review* 58 (February), 35–36.

Halaby, Charles N., and David L. Weakliem. (1993). Ownership and authority in the earnings function: Non-nested tests of alternative specifications. *American Sociological Review* 58 (February), 16–30.

Hartmann, Heidi. (1981). The unhappy marriage of Marxism and feminism: Toward a more progressive union. In Lydia Sargent (Ed.), *Women and revolution* (pp. 1–41). Boston: South End Press.

Henry, F., C. Tator, W. Mattis, and T. Rees. (2000). *The colour of democracy* (2nd ed.). Toronto: Harcourt Canada.

Hindess, B., and P.Q. Hirst. (1975). *Pre-capitalist modes of production.* London: Routledge.

Hindess, B., and P.Q. Hirst. (1977). *Modes of production and social formation*. London: Macmillan.

Hodge, Robert W., V. Kraus, and E.O. Schild. (1982). Consensus in occupational prestige research: Response to Guppy. *Social Forces* 60 (June), 1190–1196.

Hodge, Robert W., Paul M. Siegel, and Peter H. Rossi. (1964). Occupational prestige in the United States: 1925–1963. *American Journal of Sociology* 70 (November), 286–302.

Hodge, Robert W., Donald J. Treiman, and Peter H. Rossi. (1966). A comparative study of occupational prestige. In R. Bendix and S.M. Lipset (Eds.), *Class, status, and power* (2nd ed.) (pp. 309–321). New York: The Free Press.

Hopcroft, Rosemary L. (1994). The social origins of agrarian change in late medieval England. *American Journal of Sociology* 99 (May), 1559–1595.

Hout, Mike, Clem Brooks, and Jeff Manza. (1993). The persistence of classes in post-industrial societies. *International Sociology* 8 (September), 259–277.

Huber, Joan. (1990). Macro-micro links in gender stratification. *American Sociological Review* 55 (February), 1–10.

Huber, Joan (Ed.). (1991). *Macro-micro linkages in sociology*. Newbury Park, CA: Sage.

Huber, Joan, and Glenna Spitze. (1983). *Sex stratification: Children, housework, and jobs*. New York: Academic Press.

Hunter, Alfred A. (1986). *Class tells: On social inequality in Canada* (2nd ed.). Toronto: Butterworths.

Hunter, Alfred A. (1988). Formal education and initial employment: Unravelling the relationships between schooling and skills over time. *American Sociological Review* 53 (February), 753–765.

Hurst, Charles E. (1995). *Social inequality: Forms, causes, and consequences* (2nd ed.). Boston: Allyn and Bacon.

Inkeles, Alex, and Peter Rossi. (1956). National comparison of occupational prestige. *American Journal of Sociology* 61 (January), 329–339.

Inkeles, Alex, and David Horton Smith. (1974). *Becoming modern*. Cambridge, MA: Harvard University Press.

Innis, Harold A. (1956). *The fur trade in Canada*. Toronto: University of Toronto Press.

Jackman, Mary. (1994). *The velvet glove: Paternalism and conflict in gender, class, and race*. Berkeley and Los Angeles: University of California Press.

Jasso, G., and P.H. Rossi. (1977). Distributive justice and earned income. *American Sociological Review* 42 (August), 639–651.

Jessop, Bob. (1982). *The capitalist state*. Oxford: Martin Robertson.

Jessop, Bob. (1985). *Nicos Poulantzas: Marxist theory and political strategy.* New York: St. Martin's Press.

Jessop, Bob. (1991). On the originality, legacy, and actuality of Nicos Poulantzas. *Studies in Political Economy* 34 (Spring), 75–107.

Johnson, Harry M. (1960). *Sociology: A systematic introduction.* New York: Harcourt, Brace and World.

Johnson, Leo. (1979). Income disparity and the structure of earnings in Canada, 1946–74. In J.E. Curtis and W.G. Scott (Eds.), *Social stratification: Canada* (2nd ed.) (pp. 141–157). Scarborough: Prentice-Hall Canada.

Johnson, Miriam. (1993). Functionalism and feminism: Is estrangement necessary? In Paula England (Ed.), *Theory on gender/feminism on theory* (pp. 115–130). New York: Aldine de Gruyter.

Johnston, William, and Michael Ornstein. (1982). Class, work, and politics. *Canadian Review of Sociology and Anthropology* 19 (2) (May), 196–214.

Jones, Robert A. (1994). Ambivalent Cartesians: Durkheim, Montesquieu, and method. *American Journal of Sociology* 100 (1) (July), 1–39.

Kallen, Horace M. (1931). Functionalism. In E. Seligman (Ed.), *Encyclopedia of the social sciences* (Vol. 6, pp. 523–526). New York: Macmillan and The Free Press.

Kamolnick, Paul. (1988). *Classes: A Marxist critique.* Dix Hills, NY: General Hall.

Käsler, P. (1988). *Max Weber. An introduction to his life and work.* Cambridge: Polity Press.

Kelley, Jonathan, and M.D.R. Evans. (1993). The legitimation of inequality: Occupational earnings in nine nations. *American Journal of Sociology* 99 (1) (July), 75–126.

Kivisto, Peter (Ed.). (2003). *Social theory. Roots and branches* (4th ed.). Los Angeles, CA: Roxbury.

Kluegel, James, and Lawrence Bobo. (1993). Opposition to race-targeting. *American Sociological Review* 58 (August), 443–464.

Kolko, Gabriel. (1962). *Wealth and power in America.* New York: Praeger.

Krymkowski, Daniel. (2000). The puzzle of Lenski's curve. *Rationality and Society* 12 (1) (February), 25–38.

Lachmann, Richard. (1990). Class formation without class struggle: An elite conflict theory of the transition to capitalism. *American Sociological Review* 55 (June), 398–414.

Lambert, Ronald D., James Curtis, Steven Brown, and Barry Kay. (1986). Canadians' beliefs about differences between social classes. *Canadian Journal of Sociology* 11 (4) (Winter), 379–399.

Lane, David. (1982). *The end of social inequality? Class, status, and power under state socialism.* Winchester, MA: Allen and Unwin.

Larana, Enrique, Hank Johnston, and Joseph Gusfield (Eds.). (1994). *New social movements: From ideology to identity*. Philadelphia: Temple University Press.

Laxer, Gordon. (2004). Democracy and global capitalism. In J. Curtis, E. Grabb, and N. Guppy (Eds.), *Social inequality in Canada: Patterns, problems, policies* (4th ed.) (pp. 31–37). Toronto: Pearson Education Canada.

Lehmann, Jennifer M. (1994). *Durkheim and women*. Lincoln: University of Nebraska Press.

Lehmann, Jennifer M. (1995a). Durkheim's theories of deviance and suicide: A feminist reconsideration. *American Journal of Sociology* 100 (January), 904–930.

Lehmann, Jennifer M. (1995b). The question of caste in modern society: Durkheim's contradictory theories of race, class, and sex. *American Sociological Review* 60 (August), 566–585.

Lemert, Charles. (1995). *Sociology after the crisis*. Boulder, CO: Westview Press.

Lemke, Christiane, and Gary Marks (Eds.). (1992). *The crisis of socialism in Europe*. Durham, NC: Duke University Press.

Lenin, V.I. (1917). The state and revolution. In *Selected works*. London: Lawrence and Wishart, 1969.

Lenski, Gerhard E. (1966). *Power and privilege: A theory of social stratification*. New York: McGraw-Hill.

Lenski, Gerhard E. (1980). In praise of Mosca and Michels. *Mid-American Review of Sociology* 5 (2), 1–12.

Lenski, Gerhard E. (1988). Rethinking macrosociological theory. *American Sociological Review* 53 (2) (April), 163–171.

Lenski, Gerhard E. (1996). Evolutionary theory and societal transformation in post-communist Europe. *Czech Sociological Review* 4 (2) (Fall), 149–156.

Lenski, Gerhard E. (2005). *Evolutionary-ecological theory: Principles and applications*. Boulder, CO: Paradigm Publishers.

Levine, Rhonda F., and Jerry Lembcke (Eds.). (1987). *Recapturing Marxism: An appraisal of recent trends in sociological theory*. New York: Praeger.

Lévi-Strauss, Claude. (1968). *Structural anthropology*. London: Allen Lane.

Levy, Marion J., Jr. (1968). Structural-functional analysis. In D.L. Sills (Ed.), *International encyclopedia of the social sciences* (Vol. 6, pp. 21–29). New York: Macmillan and The Free Press.

Li, Peter S. (1988). *Ethnic inequality in a class society*. Toronto: Wall and Thompson.

Li, Peter S. (1992). Race and gender as bases of class fractions and their effects on earnings. *Canadian Review of Sociology and Anthropology* 29 (November), 488–510.

Lips, Hilary M. (1991). *Women, men, and power*. Mountain View, CA: Mayfield Publishing Company.

Lipset, S.M. (1967). Values, education, and entrepreneurship. In S.M. Lipset and A. Solari (Eds.), *Elites in Latin America* (pp. 3–60). New York: Oxford University Press.

Lipset, S.M., and Reinhard Bendix. (1963). *Social mobility in industrial society*. Berkeley: University of California Press.

Lockwood, David. (1956). Some remarks on 'The social system.' *British Journal of Sociology* 7 (June), 134–146.

Löwith, Karl. (1982). *Max Weber and Karl Marx*. London: George Allen and Unwin.

Lukes, S.M. (1973). *Émile Durkheim: His life and work*. Harmondsworth: Penguin Books.

Lukes, S.M. (1974). *Power: A radical view*. London: Macmillan.

Lukes, S.M. (1978). Power and authority. In T. Bottomore and R. Nisbet (Eds.), *A history of sociological analysis* (pp. 633–676). New York: Basic Books.

Luxton, Meg. (1980). *More than a labour of love*. Toronto: The Women's Press.

Mackie, Marlene. (1991). *Gender relations in Canada*. Toronto: Butterworths.

Malinowski, Bronislaw. (1926). *Crime and custom in savage society*. London: Routledge.

Malinowski, Bronislaw. (1929). *The sexual life of savages in Northwest Melanesia*. London: Routledge.

Mann, Michael. (1986). *The sources of social power* (Vol. 1). Cambridge: Cambridge University Press.

Mann, Michael. (1993). *The sources of social power* (Vol. 2). Cambridge: Cambridge University Press.

Manza, Jeff. (1992). Classes, status groups, and social classes: A critique of neo-Weberian social theory. In Ben Agger (Ed.), *Current perspectives in social theory* (Vol. 12, pp. 275–302). Greenwich, CT: JAI Press.

Marcuse, Herbert. (1971). Industrialization and capitalism. In O. Stammer (Ed.), *Max Weber and sociology today* (pp. 133–151). New York: Harper and Row.

Marx, Gary, and Douglas McAdam. (1994). *Collective behavior and social movements*. Englewood Cliffs: Prentice-Hall.

Marx, Karl. (1843). Contribution to the critique of Hegel's philosophy of law. In *Marx Engels collected works* (Vol. 3). New York: International Publishers, 1975.

Marx, Karl. (1844). Economic and philosophic manuscripts of 1844. In *Marx Engels collected works* (Vol. 3). New York: International Publishers, 1975.

Marx, Karl. (1847). The poverty of philosophy. In *Marx Engels collected works* (Vol. 6). New York: International Publishers, 1976.

Marx, Karl. (1858). *Grundrisse. Foundations of the critique of political economy.* Harmondsworth: Penguin, 1973.

Marx, Karl. (1859). A contribution to the critique of political economy. Excerpt in H. Selsam, D. Goldway, and H. Martel, (Eds.), *Dynamics of social change.* New York: International Publishers, 1970.

Marx, Karl. (1862). *Theories of surplus value* (Vol. 2). Moscow: Progress Publishers, 1968.

Marx, Karl. (1863). *Theories of surplus value* (Vol. 3). Moscow: Progress Publishers, 1971.

Marx, Karl. (1867). *Capital* (Vol. 1). New York: International Publishers, 1967.

Marx, Karl. (1875). *Critique of the Gotha program.* New York: International Publishers, 1938.

Marx, Karl. (1894). *Capital* (Vol. 3). New York: International Publishers, 1967.

Marx, Karl, and Friedrich Engels. (1846). The German ideology. In *Marx Engels collected works* (Vol. 5). New York: International Publishers, 1976.

Marx, Karl, and Friedrich Engels. (1848). *The communist manifesto.* New York: Washington Square Press, 1970.

Matthews, Ralph. (1983). *The creation of regional dependency.* Toronto: University of Toronto Press.

Mayer, Thomas F. (1994). Analytical Marxism. *Contemporary social theory* (Vol. 1). New York: Sage.

McLellan, David. (1971). *The thought of Karl Marx.* New York: Harper and Row.

McLellan, David. (1973). *Karl Marx: His life and thought.* London: Macmillan.

McMullin, Julie. (2004). *Understanding social inequality: Intersections of class, age, gender, ethnicity, and race in Canada.* Don Mills: Oxford University Press.

Meiksins, Peter. (1987). New classes and old theories: The impasse of contemporary class analysis. In R.F. Levine and J. Lembcke (Eds.), *Recapturing Marxism: An appraisal of recent trends in sociological theory* (pp. 37–63). New York: Praeger.

Merton, Robert K. (1949). *Social theory and social structure.* New York: The Free Press.

Michels, Robert. (1915). *Political parties. A sociological study of the oligarchical tendencies of modern democracy.* New York: The Free Press, 1962.

Milanovic, Branko. (2005). *Worlds apart: Measuring international and global inequality.* Princeton, NJ: Princeton University Press.

Miliband, Ralph. (1969). *The state in capitalist society*. London: Weidenfeld and Nicolson.

Miliband, Ralph. (1987). Classes. In Anthony Giddens and Jonathan Turner (Eds.), *Sociological theory today* (pp. 325–346). Stanford: Stanford University Press.

Milkman, Ruth. (1987). *Gender at work*. Urbana and Chicago: University of Illinois Press.

Millet, Kate. (1969). *Sexual politics*. Garden City: Doubleday.

Mills, C. Wright. (1951). *White collar*. New York: Oxford University Press.

Mills, C. Wright. (1956). *The power elite*. New York: Oxford University Press.

Mills, C. Wright. (1959). *The sociological imagination*. New York: Oxford University Press.

Morris, Martina, Annette Bernhardt, and Mark Handcock. (1994). Economic inequality: New methods for new trends. *American Sociological Review* 59 (April), 205–219.

Mouzelis, N. (1995). *Sociological theory: What went wrong? Diagnoses and remedies*. London: Routledge.

Münch, Richard. (1982). Talcott Parsons and the theory of action. II. The continuity of the development. *American Journal of Sociology* 87 (January), 771–826.

Murgatroyd, Linda. (1989). Only half the story: Some blinkering effects of 'malestream' sociology. In D. Held and J.B. Thompson (Eds.), *Social theory of modern societies: Anthony Giddens and his critics* (pp. 147–161). Cambridge: Cambridge University Press.

Murphy, Raymond. (1988). *Social closure: The theory of monopolization and exclusion*. Oxford: Clarendon Press.

Myles, John. (1988). The expanding middle: Some Canadian evidence on the deskilling debate. *Canadian Review of Sociology and Anthropology* 25 (August), 335–364.

Nelson, Adie. (2005). *Gender in Canada* (3rd ed.). Toronto: Pearson Education Canada.

Nielsen, Francois. (1994). Income inequality and industrial development: Dualism revisited. *American Sociological Review* 59 (October), 654–677.

Nisbet, Robert A. (1959). The decline and fall of social class. *Pacific Sociological Review* 2 (Spring), 11–17.

Noel, Donald. (1968). A theory of the origin of ethnic stratification. *Social Problems* 16, 157–172.

Nolan, Patrick, and Gerhard E. Lenski. (1996). Trajectories, ideology, and societal development. *Sociological Perspectives* 39 (1) (Spring), 23–38.

North, Cecil C., and Paul K. Hatt. (1947). Jobs and occupations: A popular evaluation. *Opinion News* (September).

Nosanchuk, T.A. (1972). A note on the use of the correlation coefficient for assessing the similarity of occupational rankings. *Canadian Review of Sociology and Anthropology* 9 (November), 357–365.

Oakley, Ann. (1974). *The sociology of housework*. Oxford: Martin Robertson.

Offe, Claus. (1974). Structural problems of the capitalist state. *German Political Studies* 1.

Offe, Claus. (1984). *Contradictions of the welfare state*. Cambridge, MA: MIT Press.

Offe, Claus, and Volker Ronge. (1975). Theses on the theory of the state. *New German Critique* 6, 139–147.

Olzak, Susan. (1992). *The dynamics of ethnic competition and conflict*. Stanford: Stanford University Press.

Olzak, Susan, and Joane Nagel (Eds.). (1986). *Competitive ethnic relations*. Orlando: Academic Press.

Pakulski, Jan. (2005). Foundations of a post-class analysis. In E.O. Wright (Ed.), *Approaches to class analysis* (pp. 152–179). Cambridge: Cambridge University Press.

Parkin, Frank. (1972). *Class inequality and political order*. London: Paladin.

Parkin, Frank. (1978). Social stratification. In T. Bottomore and R. Nisbet (Eds.), *A history of sociological analysis* (pp. 599–632). New York: Basic Books.

Parkin, Frank. (1979). *Marxism and class theory: A bourgeois critique*. London: Tavistock.

Parkin, Frank. (1980). Reply to Giddens. *Theory and Society* 9 (November), 891–894.

Parkin, Frank. (1983). Strategies of social closure in class formation. *Soziale Welt*, Supplement 2, 121–135.

Parkin, Frank. (1992). *Durkheim*. Oxford: Oxford University Press.

Parsons, Talcott. (1937). *The structure of social action* (Vol. 1). New York: The Free Press.

Parsons, Talcott. (1940). An analytical approach to the theory of social stratification. In T. Parsons, *Essays in sociological theory* (pp. 69–88). New York: The Free Press, 1964.

Parsons, Talcott. (1947). *Max Weber: The theory of social and economic organization*. New York: The Free Press.

Parsons, Talcott. (1951). *The social system*. New York: The Free Press.

Parsons, Talcott. (1953). A revised analytical approach to the theory of social stratification. In T. Parsons, *Essays in sociological theory* (pp. 386–439). New York: The Free Press, 1964.

Parsons, Talcott. (1954). *Essays in sociological theory* (Rev. ed.). Glencoe, IL: The Free Press.

Parsons, Talcott. (1966). On the concept of political power. In R. Bendix and S.M. Lipset (Eds.), *Class, status, and power* (2nd ed.) (pp. 240–265). New York: The Free Press.

Patocka, Jakub. (1998). From Marxist revolution to technological revolution. *Czech Sociological Review* 6 (1) (Spring), 115–121.

Peyre, Henri. (1960). Durkheim: The man, his time, and his intellectual background. In Kurt Wolff (Ed.), *Émile Durkheim* (pp. 3–31). Columbus: Ohio State University Press.

Picot, Garnett, John Myles, and Ted Wannell. (1990). Good jobs/bad jobs and the declining middle: 1967–1986. Research Paper No. 28, Analytical Studies Branch, Statistics Canada.

Pincus, Fred L., and Howard J. Ehrlich (Eds.). (1994). *Race and ethnic conflict.* Boulder, CO: Westview Press.

Pineo, Peter, John Porter, and Hugh McRoberts. (1977). The 1971 census and the socioeconomic classification of occupations. *Canadian Review of Sociology and Anthropology* 14 (February), 91–102.

Porter, John. (1965). *The vertical mosaic: An analysis of social class and power in Canada.* Toronto: University of Toronto Press.

Portis, Edward Bryan. (1986). *Max Weber and political commitment: Science, politics and personality.* Philadelphia: Temple University Press.

Poulantzas, Nicos. (1973a). *Political power and social classes.* London: New Left Books.

Poulantzas, Nicos. (1973b). On social classes. *New Left Review* 78, 27–54.

Poulantzas, Nicos. (1975). *Classes in contemporary capitalism.* London: New Left Books.

Poulantzas, Nicos. (1978). *State, power, socialism.* London: New Left Books.

Przeworski, Adam. (1985). *Capitalism and social democracy.* Cambridge: Cambridge University Press.

Radcliffe-Brown, A.R. (1922). *The Andaman Islanders.* Cambridge: Cambridge University Press.

Radcliffe-Brown, A.R. (1935). On the concept of function in social science. *American Anthropologist* 37 (July–September), 395–402.

Radcliffe-Brown, A.R. (1948). *A natural science of society.* New York: The Free Press.

Radcliffe-Brown, A.R. (1952). *Structure and function in primitive society: Essays and addresses.* London: Cohen and West.

Ramcharan, Subhas. (1982). *Racism: Non-whites in Canada.* Toronto: Butterworths.

Reskin, Barbara, and Heidi Hartmann. (1986). *Women's work, men's work: Sex segregation on the job.* Washington: National Academy Press.

Reskin, Barbara, and Irene Padavic. (1994). *Women and men at work.* Thousand Oaks, CA: Pine Forge Press.

Reskin, Barbara, and Patricia Roos. (1990). *Job queues, gender queues.* Philadelphia: Temple University Press.

Richardson, Laurel. (1988). *The dynamics of sex and gender: A sociological perspective.* New York: Harper and Row.

Riesman, David, N. Glazer, and R. Denney. (1953). *The lonely crowd.* New York: Doubleday.

Rinehart, James. (2006). *The tyranny of work* (5th ed.). Toronto: Thomson Nelson.

Ritzer, George. (1988). *Contemporary sociological theory* (2nd ed.). New York: Alfred Knopf.

Ritzer, George, and Douglas Goodman. (2003). *Classical sociological theory* (4th ed.). Burr Ridge, IL: McGraw-Hill Higher Education.

Robinson, Robert V., and Wendell Bell. (1978). Equality, success, and social justice in England and the United States. *American Sociological Review* 43 (April), 125–143.

Roemer, John. (1982). *A general theory of exploitation and class.* Cambridge, MA: Harvard University Press.

Rothman, Robert A. (1993). *Inequality and stratification: Class, color, and gender* (2nd ed.). Englewood Cliffs: Prentice-Hall.

Runciman, W.G. (Ed.). (1978). *Max Weber: Selections in translation.* Cambridge: Cambridge University Press.

Runciman, W.G. (1989). *A treatise on social theory: Vol. 2. Substantive social theory.* Cambridge: Cambridge University Press.

Salomon, Albert. (1945). German sociology. In Georges Gurvitch and Wilbert E. Moore (Eds.), *Twentieth century sociology.* New York: Philosophical Library.

Sanday, Peggy Reeves. (1981). *Female power and male dominance: On the origins of sexual inequality.* Cambridge: Cambridge University Press.

Satzewich, Vic (Ed.). (1998). *Racism and social inequality in Canada.* Toronto: Thompson Educational Publishing.

Saunders, Eileen. (1999). Theoretical approaches to the study of women. In James Curtis, Edward Grabb, and Neil Guppy (Eds.), *Social inequality in Canada: Patterns, problems, policies* (3rd ed.) (pp. 168–185). Scarborough: Prentice-Hall Canada.

Sayer, Derek. (1991). *Capitalism and modernity: An excursus on Marx and Weber.* London: Routledge.

Schacht, Richard. (1970). *Alienation.* Garden City: Doubleday.

Scott, Alison MacEwen. (1986). Industrialization, gender segregation, and stratification theory. In Rosemary Crompton and Michael Mann (Eds.), *Gender and stratification* (pp. 154–183). Cambridge: Polity Press.

Scott, John. (1979). *Corporations, classes, and capitalism.* London: Hutchinson.

Scott, John. (1995). *Sociological theory: Contemporary debates.* Brookfield, VT: Edward Elgar Publishing Company.

Scott, John. (1996). *Stratification and power: Structures of class, status, and command.* Cambridge: Polity Press.

Selsam, Howard, David Goldway, and Harry Martel (Eds.). (1970). *Dynamics of social change.* New York: International Publishers.

Sewell, William H., Jr. (1992). A theory of structure: Duality, agency, and transformation. *American Journal of Sociology* 98 (July), 1–29.

Shelton, Beth Anne, and Ben Agger. (1993). Shotgun wedding, unhappy marriage, no-fault divorce? Rethinking the feminism–Marxism relationship. In Paula England (Ed.), *Theory on gender/feminism on theory* (pp. 25–41). New York: Aldine de Gruyter.

Shibutani, T., and K.M. Kwan. (1965). *Ethnic stratification: A comparative approach.* New York: Macmillan.

Shils, Edward, and Henry Finch (Eds.). (1949). *The methodology of the social sciences. Max Weber.* New York: The Free Press.

Siltanen, Janet. (1994). *Locating gender.* London: University College London Press.

Siltanen, Janet. (2004). Inequalities of gender and class: Charting the sea change. In J. Curtis, E. Grabb, and N. Guppy (Eds.), *Social inequality in Canada: Patterns, problems, policies* (4th ed.) (pp. 215–230). Toronto: Pearson Education Canada.

Siltanen, Janet, Jennifer Jarman, and Robert Blackburn. (1995). *Gender inequality in the labour market. Occupational concentration and segregation.* Geneva: ILO.

Singer, Peter. (1980). *Marx.* Oxford: Oxford University Press.

Sorokin, Pitirim A. (1927). *Social mobility.* New York: Harper and Row.

Sorokin, Pitirim A. (1947). *Society, culture, and personality.* New York: Harper and Row.

Stanworth, Michelle. (1984). Women and class analysis: A reply to John Goldthorpe. *Sociology* 18, 159–170.

Stasiulis, Daiva. (1990). Theorizing connections: Gender, race, ethnicity, and class. In Peter Li (Ed.), *Race and ethnic relations in Canada* (pp. 269–305). Toronto: Oxford University Press.

Stehr, Nico. (1974). Consensus and dissensus in occupational prestige. *British Journal of Sociology* 25 (December), 410–427.

Steinmetz, George, and Erik Olin Wright. (1989). The fall and rise of the petty bourgeoisie: Changing patterns of self-employment in the post-war United States. *American Journal of Sociology* 94 (5) (March), 973–1018.

Stolzman, James, and Herbert Gamberg. (1974). Marxist class analysis versus stratification analysis as general approaches to social inequality. *Berkeley Journal of Sociology* 18, 105–125.

Sydie, R.A. (1987). *Natural women, cultured men: A feminist perspective on sociological theory.* Toronto: Methuen.

Sydie, R.A. (1994). Sex and the sociological fathers. *Canadian Review of Sociology and Anthropology* 31 (May), 117–138.

Therborn, Goran. (1986). Class analysis: History and defence. In Ulf Himmelstrand (Ed.), *Sociology: From crisis to science* (Vol. 1, pp. 96–132). London: Sage.

Tilly, Charles. (1998). *Durable inequality.* Berkeley: University of California Press.

Tiryakian, Edward. (1995). Review of Jennifer Lehmann's 'Durkheim and women.' *American Journal of Sociology* 100 (March), 1375–1377.

Treiman, Donald J. (1977). *Occupational prestige in comparative perspective.* New York: Academic Press.

Tumin, Melvin M. (1953). Some principles of stratification: A critical analysis. *American Sociological Review* 18 (August), 387–393.

Turner, Bryan S. (1992). *Max Weber. From history to modernity.* London: Routledge.

Turner, Jonathan H. (1984). *Societal stratification: A theoretical analysis.* New York: Columbia University Press.

Turner, Jonathan H. (1986). The theory of structuration. *American Journal of Sociology* 86 (4) (January), 969–977.

Veenstra, Gerry. (2005). Can taste illumine class? Cultural knowledge and forms of inequality. *Canadian Journal of Sociology* 30 (3) (Summer), 247–280.

Walby, Sylvia. (1986). *Patriarchy at work.* Cambridge: Polity Press.

Walby, Sylvia. (1989). Theorising patriarchy. *Sociology* 23 (2) (May), 213–234.

Walby, Sylvia. (1990). *Theorizing patriarchy.* Oxford: Basil Blackwell.

Walby, Sylvia. (1997). *Gender transformations.* London: Routledge.

Walby, Sylvia (Ed.). (1999). *New agendas for women.* Houndmills, Basingstoke: Macmillan; New York: St. Martin's Press.

Walby, Sylvia. (2002). Feminism in a global era. *Economy and Society* 31 (4), 533–557.

Walby, Sylvia. (2005). Gender mainstreaming: Productive tensions in theory and practice. *Social Politics* 12 (Fall), 321–343.

Wallace, Ruth (Ed.). (1989). *Feminism and sociological theory.* Newbury Park: Sage.

Wallerstein, Immanuel. (1974). *The modern world-system* (Vol. 1). New York: Academic Press.

Wallerstein, Immanuel. (1979). *The capitalist world-economy.* Cambridge: Cambridge University Press.

Wallerstein, Immanuel. (1989). *The modern world-system* (Vol. 3). New York: Academic Press.

Wallerstein, Immanuel. (1995). *After liberalism.* New York: The New Press.

Wallerstein, Immanuel. (1998). Contemporary capitalist dilemmas, the social sciences, and the geopolitics of the twenty-first century. *Canadian Journal of Sociology* 23 (2/3) (Spring-Summer), 141–158.

Ward, Kathryn B. (1993). Reconceptualizing world system theory to include women. In Paula England (Ed.), *Theory on gender/feminism on theory* (pp. 43–68). New York: Aldine de Gruyter.

Weber, Marianne. (1926). *Max Weber: A biography.* (Harry Zohn, Ed. and Trans.). New York: John Wiley and Sons, 1975.

Weber, Max. (1905). *The Protestant ethic and the spirit of capitalism.* New York: Charles Scribner's Sons, 1958.

Weber, Max. (1922). *Economy and society* (Vols. 1–3). New York: Bedminster Press, 1968.

Weber, Max. (1923). *General economic history.* New York: Collier-Macmillan, 1961.

Weiss, Donald D. (1976). Marx versus Smith on the division of labor. In *Technology, the labor process, and the working class* (pp. 104–118). New York: Monthly Review Press.

Wenger, Morton G. (1980). The transmutation of Weber's Stand in American sociology and its social roots. In S. McNall and G. Howe (Eds.), *Current perspectives in social theory* (Vol. 1, pp. 357–378). Greenwich, CT: JAI Press.

Wenger, Morton G. (1987). Class closure and the historical/structural limits of the Marx-Weber convergence. In Norbert Wiley (Ed.), *The Marx-Weber debate* (pp. 43–64). Newbury Park, CA: Sage.

Wesolowski, W. (1966). Some notes on the functional theory of stratification. In R. Bendix and S.M. Lipset (Eds.), *Class, status, and power* (2nd ed.) (pp. 64–69). New York: The Free Press.

Western, Mark, and Erik Olin Wright. (1994). The permeability of class boundaries to intergenerational mobility among men in the United States, Canada, Norway, and Sweden. *American Sociological Review* 59 (August), 606–629.

Wiley, Norbert (Ed.). (1987). *The Marx-Weber debate.* Newbury Park, CA: Sage.

Williams, Robin M.Jr. (1960). *American society. A sociological interpretation.* New York: Knopf.

Wilson, John. (1993). The subject woman. In Paula England (Ed.), *Theory on gender/feminism on theory* (pp. 343–357). New York: Aldine de Gruyter.

Wilson, William Julius. (1973). *Power, racism, and privilege.* New York: Macmillan.

Wilson, William Julius. (1978). *The declining significance of race.* Chicago: University of Chicago Press.

Wilson, William Julius. (1987). *The truly disadvantaged.* Chicago: University of Chicago Press.

Wilson, William Julius. (1991). Studying inner-city social dislocations: The challenge of public agenda research. *American Sociological Review* 56 (February), 1–14.

Wilson, William Julius. (1999). *The bridge over the racial divide: Rising inequality and coalition politics.* Berkeley: University of California Press; New York: Russell Sage.

Wood, Ellen. (1986). *The retreat from class. A new 'true' socialism.* London: Verso.

Wright, Erik Olin. (1976). Class boundaries in advanced capitalism. *New Left Review* 98, 3–41.

Wright, Erik Olin. (1978). *Class, crisis, and the state.* London: New Left Books.

Wright, Erik Olin. (1979). *Class structure and income determination.* New York: Academic Press.

Wright, Erik Olin. (1980). Class and occupation. *Theory and Society* 9 (January), 177–214.

Wright, Erik Olin. (1985). *Classes.* London: Verso.

Wright, Erik Olin. (1989a). The comparative project on class structure and class consciousness: An overview. *Acta Sociologica* 32 (1), 3–22.

Wright, Erik Olin. (1989b). *The debate on classes.* London: Verso.

Wright, Erik Olin. (1993). Typologies, scales, and class analysis. *American Sociological Review* 58 (February), 31–34.

Wright, Erik Olin. (1994). *Interrogating inequality: Essays on class analysis, socialism, and Marxism.* London: Verso.

Wright, Erik Olin. (1997). *Class counts: Comparative studies in class analysis.* Cambridge: Cambridge University Press.

Wright, Erik Olin. (2000). Working-class power, capitalist class interests, and class compromise. *American Journal of Sociology* 105 (4) (January), 957–1002.

Wright, Erik Olin. (2002). The shadow of exploitation in Weber's class analysis. *American Sociological Review* 67 (6) (December), 832–853.

Wright, Erik Olin (Ed.). (2005). *Approaches to class analysis.* Cambridge: Cambridge University Press.

Wright, Erik Olin, C. Costello, D. Hachen, and J. Sprague. (1982). The American class structure. *American Sociological Review* 47 (December), 709–726.

Wright, Erik Olin, Andrew Levine, and Elliott Sober. (1992). *Reconstructing Marxism.* London: Verso.

Wright, Erik Olin, and Bill Martin. (1987). The transformation of the American class structure, 1960–1980. *American Journal of Sociology* 93 (1) (July), 1–29.

Wright, Erik Olin, and Luca Perrone. (1977). Marxist class categories and income inequality. *American Sociological Review* 42 (February), 32–55.

Wright, Erik Olin, and Joachim Singelmann. (1982). Proletarianization in the changing American class structure. *American Journal of Sociology* 88 (Supplement), S176–S209.

Wrong, Dennis. (1959). The functional theory of stratification: Some neglected considerations. *American Sociological Review* 24 (December), 772–782.

Wrong, Dennis. (1961). The over-socialized conception of man in modern sociology. *American Sociological Review* 26 (April), 183–193.

Wrong, Dennis. (1979). *Power: Its forms, bases and uses.* New York: Harper and Row.

Yanowitch, Murray. (1977). *Social and economic inequality in the Soviet Union.* London: Martin Robertson.

Zeitlin, Irving M. (1968). *Ideology and the development of sociological theory.* Englewood Cliffs: Prentice-Hall.

Zhou, Xueguang. (2005). The institutional logic of occupational prestige ranking: Reconceptualization and reanalyses. *American Journal of Sociology* 111 (July), 90–140.

Index

About the Cover

In *La Grande Jatte* Georges Seurat represents a disparate group of people at leisure in 1884. They have gathered at an island park in the semi-industrialized suburbs of Paris to fish and boat in the Seine River, read or relax in the shade, and show off their Sunday clothes.

At right is a fashionable couple of the upper class. The man is stylishly equipped with top hat, boutonnière, and monocle, sporting a walking stick in one hand and a cigar in the other, while his companion wears an exaggerated bustle (padding to thrust out the back of her skirt) and calls attention to herself with her pets—the exotic small monkey and yapping little dog whose leashes she holds. At the left, a man lounges, his bare arms, pipe, and cap signalling that he is a member of the working class, possibly the owner of the black mutt that roams nearby. Further back, at left, we see a woman in red, fishing: some critics have identified her as a prostitute or "hooker" (note the French pun on *pêcher,* to fish, and *pecher,* to sin). Under the tree sits another working-class figure, a woman wearing the traditional cape and red-ribboned cap of a nurse; her elderly patient is slumped nearby under her parasol. An amateur musician in a silly tropical pith helmet blasts his trombone in front of two hatless women, one of whom turns away. We also see soldiers, small groups of women, one nuclear family, and numerous other strollers, fishers, boaters, and loungers. All have crowded on the small island park to promenade and escape the busy Paris streets.

However, the stiff postures of the figures and the lack of interaction between them hints at uneasiness among the different classes Seurat portrays. In fact, some of the figures hardly seem to occupy the same physical space. Critics have pointed out that the mother at the centre of the picture is on the same plane as the fisherwoman at left, yet she would tower over the other if the figures were placed side by side. Similarly, the sitting man at left in top hat and cane, identified by one of the painting's first reviewers as a *calicot* (shop clerk with fashionable aspirations) would only be waist high to the stylish couple at right if he were standing. Seurat's pointillist technique (a painting method using dots of colour) was considered ultra modern and scientific in 1884. But his depiction of the stilted, oblivious weekenders offers an almost humorous portrayal of anxieties about social inequality that were emerging in the late nineteenth century.

Georges Seurat, *Un dimanche à La Grande Jatte—1884*. Painted 1884–86. Oil on canvas, 207.5 cm × 308.1 cm (81¾ in. × 121¼ in.). The Art Institute of Chicago: Helen Birch Bartlett Memorial Collection.

Source: Herbert, Robert L. (Ed.) (2004). *Seurat and the Making of* La Grande Jatte. Chicago: The Art Institute of Chicago in association with the University of California Press.